Springer Tracts in Natural Philosophy

Volume 9

Magnetoelastic Interactions

William Fuller Brown, Jr.

Springer-Verlag New York Inc. 1966

Dr. William Fuller Brown, Jr.
Professor of Electrical Engineering
University of Minnesota, Minneapolis

Printed in Germany
Library of Congress Catalog Card Number 66-26600

Title-No. 6737

Preface

The modern theory of ferromagnetic magnetization processes has from the beginning recognized the importance of magnetoelastic interactions. Most of the magnetoelastic calculations, however, have been based on the theory developed by R. Becker and others in the early 1930's. That theory has several defects; how to remedy them is the subject of this monograph.

I first became aware of the shortcomings of the traditional theory thru a critical study of electric and magnetic forces, which I undertook as a member of the Coulomb's Law Committee of the American Association of Physics Teachers. My conclusions were published in 1951 in the *American Journal of Physics*; an application of them to a problem in magnetostriction was published in 1953 in *Reviews of Modern Physics.* With the development, in 1956, of the "nucleation field" theory of micromagnetics, the need for a systematic and self-consistent theory of magnetoelastic interactions became more pressing. The traditional theory predicted that the nucleation field should differ negligibly from that of a rigid body; but my 1953 magnetostriction calculation suggested that terms omitted in that theory might be important. In the academic year 1963/64, I was finally able — thanks to a sabbatical furlough — to find the time needed for systematic development of a basic theory of magnetoelastic interactions in a ferromagnet.

In the meantime, Professor C. Truesdell had called my attention to R. Toupin's 1956 paper on the elastic dielectric, a closely related topic. Toupin's most important contribution was the use of finite-strain theory; this greatly clarifies the basic relations, even for a material in which, ultimately, a small-strain approximation will be justified. The finite-strain treatment makes obsolete the small-strain treatment of magnetizable elastic solids in Section 4.4 of my 1951 paper; the treatment in the present monograph is based squarely on finite-strain theory.

On the other hand, my 1951 treatment of the stress concept still seems to me not only to be sound and adequate, but to demonstrate an important fact that had been overlooked by earlier authors and has continued to be overlooked by later ones. That fact is that a variety of mutually incompatible formulas for the "magnetic force", which lead to mutually incompatible formulas for the force expressible in terms of "stresses", are nevertheless equivalent with respect to the total force;

and only this total force is observable. Thus the incompatibility is not physical but either metaphysical (if one regards certain propositions as postulates) or semantic (if one regards them as definitions). Failure to recognize this fact has resulted in much fruitless controversy.

In the hope of promoting wider recognition of this important fact, I have included in this monograph Sections 1.2 to 1.4, and parts of Sections 1.1 and 1.5, of my 1951 article. These appear here, with minor changes of wording and notation, as §§ 5.1—5.5. The latter part of Section 1.5 has been replaced by a new discussion, in § 5.6, which demonstrates in more detail the relation of my "stresses", based on dipole-dipole forces, to alternative "stresses" based on pole-pole or current-current forces; it also relates TOUPIN'S and TIERSTEN'S "stresses" to mine. I hope that the rewriting will render more convincing the point that I was trying to make. Incidentally, I have eliminated a casual and poorly worded remark about the symmetry of stresses "defined by any energy method" (pages 297—298 of the original paper); TOUPIN rightly objected to that remark, tho I think his interpretation of it goes beyond what I actually said and overlooks the penultimate paragraph of Section 4.5 of my paper.

My acknowledgements must begin with Professor S. L. QUIMBY, who directed my doctoral research at Columbia University in the 1930's; not only did he introduce me to magnetoelasticity, but he taught me things about electromagnetic theory that I have never forgotten, including the need to be skeptical about what one finds in textbooks. Next, I must acknowledge the encouragement and the helpful comments of Professor E. C. KEMBLE, of Harvard University, who was chairman of the COULOMB'S Law Committee. I am indebted to the University of Minnesota for the sabbatical furlough that gave me the time to do the work reported here. To Professor ALFRED SEEGER, of the Max-Planck-Institut für Metallforschung and the Technische Hochschule in Stuttgart, I am indebted for the invitation to spend my sabbatical year there, for the excellent facilities made available to me, and for the suggestion that my presentation of the results take the form of a tract in this series. I am grateful to the *American Journal of Physics* for permission to reprint parts of my 1951 article. Finally, I must express my appreciation of the cooperativeness and skill of Mrs. SHARON NELSON, who miraculously transformed my pencil draft into a legible typed manuscript, and of the invaluable help of my wife, NANCY, in the proofreading of the galleys.

Minneapolis, 29 May, 1966 WILLIAM FULLER BROWN, JR.

Table of Contents

Chapter III
The Energy Method

Chapter IV
Applications

Chapter I

Fundamental Concepts and Definitions

1. Introduction

1.1. Statement of the problem. It has been known for about a century that a ferromagnetic rod, if subjected to a magnetizing field, changes its length as well as its magnetization; and if subjected to tension, changes its magnetization as well as its length. In other words, there is interaction between magnetic and elastic processes. Other and more complicated forms of the interaction are also well known. The general term for this class of phenomena is *magnetoelastic interaction* or, more briefly, *magnetostriction.*

For the rod, an elementary thermodynamic argument predicts that if one of the effects mentioned occurs, the other must also occur. The argument, however, assumes reversibility, whereas the process of magnetizing a ferromagnetic rod is a highly irreversible one: a plot of magnetization against applied field intensity gives the well-known hysteresis loop. The application of reversible thermodynamics to the observed rod behavior is therefore a dubious procedure. But in 1928 to 1930, AKULOV [*1*] and BECKER [*1*] introduced new concepts into the theory of the magnetization process. The most important change was that attention was now directed, not at the apparent magnetization (mean moment per unit volume) of the specimen as a whole, but at the local magnetization on a microscopic scale: over distances of 10^{-5} to 10^{-2} cm, small in comparison with usual specimen sizes but still large in comparison with the lattice spacing. On this scale, the magnetization has (when the temperature is uniform) a magnitude independent of position (the *spontaneous magnetization*) but a direction that can vary from one point to another. By a combination of atomic and phenomenological methods, an expression for the free energy, as a function of parameters that describe the magnetization distribution, can be derived. This expression can then be used to find the stable equilibrium distributions. When the applied field or tension undergoes a small change, the distribution either changes reversibly to a slightly different one, or becomes unstable; in the latter case, an irreversible transition (Barkhausen jump) occurs to a new, finitely different stable equilibrium state. Thus such a theory is in principle capable of describing both reversible magnetization processes and hysteresis.

In this theory, magnetoelastic interaction played an important role; in fact, at one time (BECKER and DÖRING [*1*]) "internal stress" (a concept never very precisely defined) seemed to be the determining factor in virtually all magnetization processes. Later it was realized that other factors were also important; but magnetoelastic interactions remain one of the important ones.

The essentials of the modern theory of magnetization processes have been summarized in several books: for example, BOZORTH [*1*], BATES [*1*], KNELLER [*1*], CHIKAZUMI [*1*], and MORRISH [*1*]. Attempts to improve its rigor, by replacing "domain theory" by "micromagnetics", have been summarized by SHTRIKMAN and TREVES [*1*] and by BROWN [*11*]; in this form of the theory, still in its infancy, the body has usually been assumed rigid in order to simplify the problem.

The theory so constructed has been, on the whole, successful in accounting for experimental results. With respect to magnetoelastic interactions, however, it has certain defects that will be discussed in the next section. The topic of this monograph is the development of a theory that is free from those defects.

As far as I know, the first step toward such a theory was taken in my 1951 paper (BROWN [*5*]). The primary aim of that paper was to clarify the concept of electric or magnetic force in a polarized material; it therefore did not treat magnetostriction specifically. I applied the theory to one special problem in magnetostriction in 1953 (BROWN [*6*]); the problem was investigated further in 1960 by GERSDORF [*1*]. In all these papers the concept of strain, when it occurred, was handled by use of the small-displacement approximations usual in physical theory; in the treatment of magnetostriction, despite the smallness of the strains actually encountered experimentally, the small-displacement approximation leads to certain difficulties. A formal treatment of the analogous electric problem, based on the rigorous equations of finite-strain theory, was published by TOUPIN [*1*] in 1956. TOUPIN's treatment requires some modifications, beyond the mere replacement of electric by magnetic quantities, for the case of a ferromagnetic material, with a spontaneous magnetization caused by exchange forces. The resulting modified theory has already been presented in outline (BROWN [*13, 14*]). It will be presented more completely here. Meanwhile, TIERSTEN [*1, 2*] has published an essentially equivalent theory.

The conventional theory and its defects will be discussed in more detail in § 1.2, and the rigorous formal theory will be discussed in § 1.3. It will then be possible to state more precisely (§ 1.4) the aim of the present treatment.

1.2. The conventional theory of magnetostriction. The standard treatment of magnetoelastic interactions (BECKER and DÖRING [*1*], Chap. 21; LEE [*1*]; KNELLER [*1*], Chap. 16) is essentially the following.

Free-energy expressions are written down for a magnetizable body incapable of deformation and for an elastic body incapable of magnetization. To the sum of these is added an "interaction" term, in the form of the volume integral of an interaction energy density of phenomenologically acceptable form (*i.e.*, consistent with the symmetry of the crystal and with the concept of spontaneous magnetization). Everything else follows from this by exact or (more often) approximate minimization procedures.

The theory so constructed accounts successfully for many experimental results, but it fails to predict one observed phenomenon: the dependence of the strain on the specimen shape, in the case of a uniformly magnetized ellipsoid. This "form effect" has usually been calculated by a special theory (BECKER and DÖRING [*1*], pp. 303—305; KNELLER [*1*], pp. 228—229), in which the free-energy expression is augmented by a term that takes account of the variation of the "demagnetizing factor" of an ellipsoid with the (assumed to be uniform) strain. Since this energy is not the volume integral of an energy density, it is overlooked by the standard theory. One aim of a complete theory must be to include such energy terms in the general equations, for a body of any shape.

But the standard theory is open to other criticisms as well. It assumes, without critical investigation, that the presence of strain does not invalidate any concepts or theorems of the standard theory of an undeformable magnetizable body, and that the presence of magnetization does not invalidate any concepts or theorems of the standard theory of an unmagnetizable elastic body. What is needed is a treatment that, from the first, assumes both deformability and magnetizability.

Finally, the standard theory uses the approximations of small-displacement elasticity theory. Accepting the usual expression for the free-energy density of an elastic body as a homogeneous quadratic function of the small "strains", it assumes that the interaction energy density may be expressed, with sufficient accuracy, as a linear function of those same "strains", with magnetization-dependent coefficients (BECKER and DÖRING [*1*], pp. 132—137; KNELLER [*1*], pp. 231—237). If u_i ($i=1, 2, 3$) are the Cartesian components of the displacement, the small "strains" are such quantities as $\partial u_1/\partial x_1$ and $\partial u_2/\partial x_3 + \partial u_3/\partial x_2$ (or half this latter quantity); here x_1, x_2, x_3 are the Cartesian components of a particle of the material (in either the unstrained or the strained state, within the approximations of the small-displacement theory). Now if stress-strain relations and the like are to be obtained to the first order in the strains by differentiation of an energy function, then that energy function must be correct *to the second order*. But in finite-strain theory it is shown that $\partial u_1/\partial x_1$ etc. are valid measures of strain only *to the first order* of small quantities. An expression valid only to the first order is sufficient in the elastic energy, which is quadratic in the strains; but is it sufficient in the interaction energy, which is linear in them? This question requires investigation; and up to the point at which a conclusion is reached, the investigation must be conducted without approximation, *i.e.*, by use of finite-strain theory.

1.3. The rigorous formal theory. TOUPIN's [*1*] treatment of a deformable dielectric (see also TRUESDELL and TOUPIN [*1*]) is based on two alternate procedures. The first (his § 9) starts with the stress concept and with a formula, attributed to MAXWELL, for "the resultant electrostatic force on a region containing polarized matter"; it leads to a set of volume and surface equilibrium equations that involve the components of field intensity, of polarization, and of a tensor that he calls "the local stress". The second procedure (his § 10) starts with a "stored energy function" and a variational principle; it leads to equilibrium equations that become equivalent to the previous ones if the "local stress" components of the first procedure are equated to suitably chosen quantities encountered in the second procedure. The first procedure does not in itself require the introduction of the energy concept, and the second procedure does not in itself require the introduction of the stress concept. Only the second involves the concept of deformation or strain, which is handled rigorously by finite-strain theory.

A physicist familiar with the experimental facts and with the conventional theory of magnetostriction, but not familiar with the mathematics of two-point tensor fields in general coordinates, will have great difficulties with TOUPIN's article. The major difficulties are purely formal ones and disappear if all the equations are rewritten in Cartesian coordinates. The prescription for doing this is quite simple: replace superscripts by subscripts and g by δ. This rule at least takes care of most situations. In the resulting simplified equations, one must still know that repeated subscripts are to be summed over and that $f_{,i}$ and $f_{,A}$ mean $\partial f/\partial x_i$ and $\partial f/\partial X_A$ respectively (where f may have subscripts before the comma).

With these aids, the physicist who wishes to read TOUPIN's article should have only occasional and minor difficulties with the mathematics. He may have difficulties with the misprints: one disadvantage of the extremely concise tensor notation is that it is almost devoid of the protective redundancy of ordinary notation, in which a misprinted symbol has a high probability of producing a meaningless combination and thus being immediately detectable. He may also have some difficulties with the fundamental concepts and postulates; of these I shall have more to say in later sections. But he should have no difficulties with the finite strain theory: it is quite simple, and logically much more satisfactory than the continual throwing away of higher-order terms that is characteristic of elasticity theory as normally discussed by physicists.

TIERSTEN'S [*1, 2*] papers are already in Cartesian coordinates and therefore do not require a prescription for translation.

1.4. The aim of this treatment. The purpose of this monograph is to construct, from first principles, a rigorous phenomenological theory of magnetoelastic interactions, with particular attention to ferromagnetic materials; to derive from it an approximate small-displacement theory, in which the approximations made are explicitly stated; and finally, on the basis of this theory, to criticize and emend the conventional theory.

The monograph is directed primarily at physicists interested in magnetostriction. If some members of the Society for Natural Philosophy read it and find something of value in it, I shall be happy; but this is not its major objective.

Because of the limited objective, I shall be content with something less than complete generality. For example, one can in a rigid body consider a free-energy density that is an arbitrary function of the direction cosines of the spontaneous magnetization and of their spatial derivatives of all orders; but in the applications to known ferromagnetic materials, no derivatives higher than the first are needed, and a quadratic function of them is sufficient, and therefore I shall without apology go over to this special case when the equations of the general case promise to become unpleasantly long. With strains also present, I shall be satisfied to examine the effect of the strains on the terms just discussed, rather than to consider an arbitrary function of the direction cosines, their spatial derivatives, and the strains.

Considerable attention will be given to the fundamental concepts, such as stress. The conventional theory is especially vulnerable at this point: concepts and theorems derived for an unmagnetized elastic material have been applied to a magnetized material quite naively, without investigation of their adequacy under the new conditions. But TOUPIN'S and TIERSTEN'S treatments are in my opinion also open to criticism in this respect: TOUPIN, for example, did not inquire under what conditions his basic force formula (the one he attributed to MAXWELL) had been derived, and whether those conditions were satisfied in the new situation to which they were being applied.

It has been fashionable lately, in developing a theory of any set of physical phenomena, to base the theory on a certain set of postulates, and to say that the ultimate justification for the postulates is the ability of the theory to produce formulas consistent with experiment.

I do not share the present enthusiasm for this postulational approach, and my unenthusiasm is a result of the very investigation (BROWN [*5*]) that forms the foundation of the theory to be presented here. That investigation revealed (pp. 347—349) specific instances of the following defect of the postulational approach. I suppose that the postulates are mutually consistent; otherwise an arbitrary proposition could be "proved" within the framework of the theory (see, for example, HILBERT and ACKERMANN [*1*], pp. 26 and 38). There still remains the following danger: some of the postulates may be superfluous, in the sense that the final formulas, relating observable quantities to each other, are not dependent on them. In that case it is possible for two theorists to develop two theories, completely equivalent with respect to relations among observable quantities, but mutually contradictory with respect to some of the superfluous postulates. The two theorists then can and often do argue until death (literally: see BROWN [*6*], p. 132) about who is right; actually both are right, if we allow a theory to include superfluous postulates that do not affect physically testable relations, and both are wrong if we forbid it to do so. It would, I think, be too restrictive to do the latter; but if superfluous postulates are allowed, they should be identified and labeled, and time should not be wasted in attacking or defending them — their inclusion or exclusion is a matter not of correctness, but of convenience and taste.

My own taste is for the avoidance of physically superfluous postulates if they can be conveniently avoided. I shall therefore not ask or try to answer such questions as: What is "the" true magnetic field intensity inside a magnetized body? What is "the" true magnetic force on a part of such a body? What are "the" stresses in such a body? Such questions need not be asked or answered in order to derive formulas for observable quantities, and therefore they will not be asked or answered in this monograph. What can be done instead will develop as the discussion proceeds.

TOUPIN ([*1*], p. 871) wrote: "The self field of a polarized dielectric will be our major concern." Accordingly, he used LORENTZ's microscopic theory ([*1*], p. 872) "to motivate an independent hypothesis concerning the electric self field of a continuous elastic dielectric medium". In the treatment to be presented here, such concepts will not be introduced; they are physically superfluous. The proof of that statement will be the derivation of all of the needed physical relations (*i.e.*, relations among observable quantities) without them.

1.5. Units and notation. The units used here are *generalized Gaussian;* that is, the equations contain a constant γ such that they reduce to the Gaussian equations when γ is set equal to 4π. The constant γ may, however, be given other values. If it is set equal to 1, the equations reduce to those of the Lorentz-Heaviside system. Both the Gaussian and the Lorentz-Heaviside systems are centimeter-gram-second systems in which charge is arbitrarily assigned mechanical dimensions. Those who prefer to use meter-kilogram-second mechanical units, to measure charge in coulombs, and to keep charge dimensionally independent of mechanical quantities may do so by assigning $\gamma/4\pi c^2$ the value 10^{-7} newton/amp^2; here c is the speed of electromagnetic waves, roughly 3×10^8 m/sec. This *Gaussian mks* system in my opinion solves the problem of combining mks and practical electrical units more satisfactorily than does the "mks rationalized" (Giorgi) system. Devotees of the Giorgi system will not be happy with my units; but I can assure them that the unhappiness that my system inflicts on them will be no greater than the unhappiness that their system over the last thirty years has inflicted on me. Fortunately there are few Giorgi enthusiasts among ferromagneticians interested in magnetostriction. Ferromagneticians usually use cgs electromagnetic units; and all they need to remember is that the electromagnetic current I and electromotive force $\mathscr{E}$ are the Gaussian I/c and $c\mathscr{E}$ — everything else, within the scope of this monograph, is the same in electromagnetic and in Gaussian.

All equations that can be conveniently written in vector symbols will normally be so written. Thus the divergence theorem will be written

$$\int \nabla \cdot \boldsymbol{v}\, d\tau = \int \boldsymbol{n} \cdot \boldsymbol{v}\, dS, \tag{1.1}$$

where $d\tau$ is a volume element and dS a surface element. The symbols ∇, $\nabla \cdot$, and $\nabla \times$ will be used, rather than grad, div, and curl, and the notations $\boldsymbol{u} \cdot \boldsymbol{v}$ and $\boldsymbol{u} \times \boldsymbol{v}$ will be used for the scalar and vector products. More specific items of notation will be explained as they occur.

When the vector notation becomes an impediment rather than a help, it will be abandoned in favor of Cartesian tensor notation. In this notation, Eq. (1.1) becomes

$$\int \frac{\partial v_i}{\partial x_i}\, d\tau = \int n_i v_i\, dS; \tag{1.2}$$

in each member, summation over any repeated subscript is understood, and the values 1, 2, 3 summed over correspond to the three Cartesian components. Thus the volume integrand in the left member of Eq. (1.2) is, in full, $\partial v_1/\partial x_1 + \partial v_2/\partial x_2 + \partial v_3/\partial x_3$, and the surface integrand in the right member is $n_1 v_1 + n_2 v_2 + n_3 v_3$. A still more compact way of writing Eq. (1.2) is

$$\int v_{i,i}\, d\tau = \int v_i n_i\, dS; \tag{1.3}$$

the subscript i preceded by a comma, after v_i, is equivalent to $(\partial/\partial x_i)$ before it.

It is sometimes necessary to treat separately the symmetric and antisymmetric parts of a quantity φ_{ij} that has two subscripts. These will be abbreviated as follows:

$$\left.\begin{aligned}\varphi_{(ij)} &\equiv \tfrac{1}{2}(\varphi_{ij}+\varphi_{ji}),\\ \varphi_{[ij]} &\equiv \tfrac{1}{2}(\varphi_{ij}-\varphi_{ji}).\end{aligned}\right\} \tag{1.4}$$

An example from small-displacement elasticity theory is the separation of the displacement gradient $u_{i,j}$ into its symmetric part (a component of strain) $u_{(i,j)} = \frac{1}{2}(\partial u_i/\partial x_j + \partial u_j/\partial x_i)$ and its antisymmetric part (a component of rotation) $u_{[i,j]} = \frac{1}{2}(\partial u_i/\partial x_j - \partial u_j/\partial x_i)$. The conciseness of the tensor notation — bought, of course, at the cost of easy readability and error-correcting redundancy — is well illustrated by this example. To exploit it, however, the two-subscript notation must be used; and this means that when one goes over from the six quantities $u_{(1,1)}$, $u_{(2,2)}$, $u_{(3,3)}$, $u_{(2,3)}$, $u_{(3,1)}$, $u_{(1,2)}$, which are convenient in volume-to-surface transformations, to the six quantities e_{xx}, e_{yy}, e_{zz}, e_{yz}, e_{zx}, e_{xy} (LOVE [*1*]) or x_1, x_2, x_3, x_4, x_5, x_6 (VOIGT [*1*], p. 563), which are convenient in thermodynamic arguments, one must remember that whereas $e_{xx} = x_1 = \partial u_1/\partial x_1 = u_{(1,1)}$, $e_{yz} = x_4 = \partial u_2/\partial x_3 + \partial u_3/\partial x_2 = 2u_{(2,3)}$.

Usually a quantity represented by a letter with two subscripts, such as t_{ij}, will have the properties of a tensor; that is, it will transform in a rotation of coordinate axes as does $u_i v_j$, where $\boldsymbol{u}$ and $\boldsymbol{v}$ are vectors. There will, however, be exceptions, which will be noted; it does not seem desirable to complicate the notation by introducing a special symbolism in such cases.

Other details of notation will be explained as the need arises.

2. Magnetostatic Fundamentals

2.1. Methods of approach; poles and currents. Classical magnetostatic theory undertakes to correlate a large body of experimental facts. These facts relate to forces of three kinds: forces exerted by current circuits on current circuits, forces exerted by magnetized bodies on magnetized bodies, and forces exerted by current circuits on magnetized bodies or *vice versa*. A mathematically self-consistent theory, compatible with all the experimental facts, can be developed in any of a number of different ways. What appears as a postulate in one form of the theory may appear as a theorem, derivable from postulates, in another; and as in elementary Euclidean plane geometry, a given set of theorems can be deduced from a given set of postulates in various orders. If certain precautions are observed, all these various forms of the theory are equivalent mathematically and also equivalent physically, in the sense that they lead to the same predictions with regard to observable quantities. The necessary precautions are: certain semiconvergent integrals (that is, integrals that have finite but nonunique values, dependent on the order in which two limiting processes are carried out) must be either handled with appropriate care or avoided altogether; a quantity that occurs in one theory must not be assumed to have any simple relation to a quantity in another theory just because they have been given the same name (such as "magnetic field intensity"); and a theoretical quantity must not be assumed to have direct physical significance, in the sense that it represents an experimentally observable quantity, just because it has been given a name that violates no rule of English syntax (for example, the

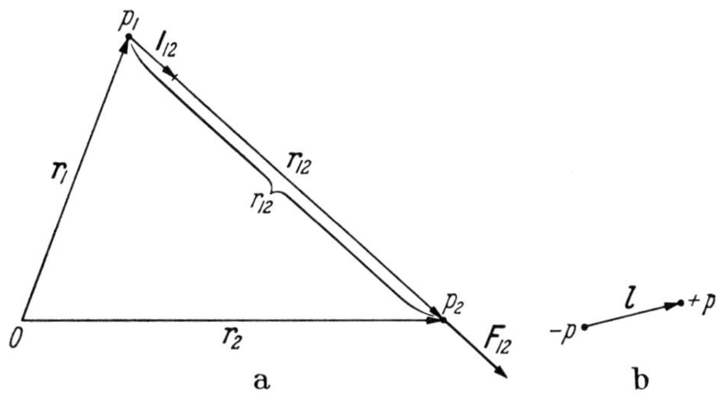

Fig. 1 (a) and (b). Notation for poles and dipoles. (a) Notation for Eq. (2.1), COULOMB's law for the force $\boldsymbol{F}_{12}$ exerted by pole p_1 on pole p_2. (b) Definition of a dipole: $\boldsymbol{l}\to 0$, $p\to\infty$, with $p\,\boldsymbol{l}=$ const $=\boldsymbol{m}$

name "field intensity at a point in the magnetized material"). Failure to observe these precautions has generated considerable confusion not only among undergraduate students, but among experts.

For a long time the theory remained mathematically simple because it began with magnet-magnet forces, basing their description on the point-pole concept; brought in magnet-current forces considerably later; and proceeded to current-current forces last of all. A point pole of strength p is defined by the law of mutual force between two such, namely COULOMB's law: the force exerted by pole p_1, at position $\boldsymbol{r}_1$, on pole p_2, at position $\boldsymbol{r}_2$, is [see Fig. 1 (a)]

$$\boldsymbol{F}_{12}=\frac{\gamma}{4\pi}\,p_1\,p_2\,\frac{\mathbf{1}_{12}}{r_{12}^2}\,, \tag{2.1}$$

where $\boldsymbol{r}_{12}=\boldsymbol{r}_2-\boldsymbol{r}_1$, r_{12} is the magnitude $|\boldsymbol{r}_{12}|$ of $\boldsymbol{r}_{12}$, and $\mathbf{1}_{12}\equiv\boldsymbol{r}_{12}/r_{12}$ is the unit vector along $\boldsymbol{r}_{12}$. Whether isolated poles exist in nature is still a subject of theoretical speculation and experimental investigation (DEVONS [1]) and is irrelevant to our topic; the role of the pole in classical magnetostatic theory is that of a convenient theoretical concept. From it we derive the *dipole* [see Fig. 1 (b)] by placing poles of strengths $+p$ and $-p$ at positions $\boldsymbol{r}+\frac{1}{2}\boldsymbol{l}$ and $\boldsymbol{r}-\frac{1}{2}\boldsymbol{l}$ respectively, and then letting l approach zero under the constraint $p\boldsymbol{l} = \text{const} \equiv \boldsymbol{m}$, the *magnetic moment* of the dipole. The forces exerted by magnetic bodies on magnetic bodies can then be described satisfactorily by supposing that the bodies carry distributions of dipoles. Then if we examine the forces exerted by magnetic bodies on current circuits and *vice versa*, we find that a small circuit, current I flowing around vector area $\boldsymbol{S} \equiv \boldsymbol{n} S$ ($\boldsymbol{n}$ = unit normal), acts like a dipole of moment $\boldsymbol{m} = I\boldsymbol{S}/c$ [see Fig. 2(a)]. By putting many small circuits side by side, so that the currents along their shared boundary segments cancel [see Fig. 2(b)], we find that a finite circuit acts like a distribution of dipoles over a

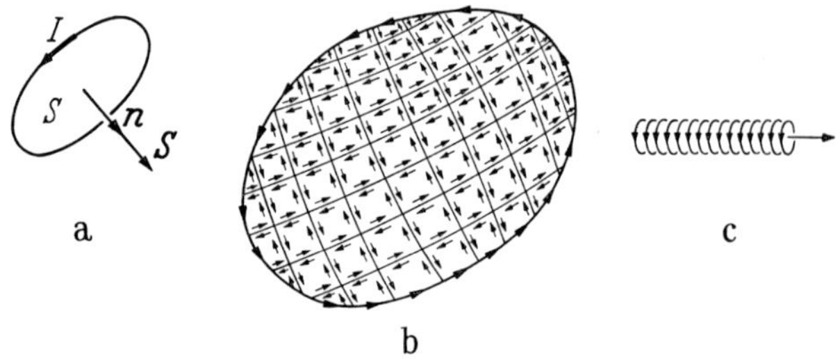

Fig. 2(a)—(c). Equivalence of currents to dipoles. (a) A small circuit: moment $\boldsymbol{m} = I\boldsymbol{S}/c$. (b) A finite circuit can be built up from small circuits side by side. (c) A long thin solenoid can be built up from small circuits one in front of another

(somewhat arbitrarily chosen) surface bounded by the circuit, with moment elements $I\,d\boldsymbol{S}/c = (I\boldsymbol{n}/c)\,dS$; by putting many small circuits one in front of another [Fig. 2(c)], we reverse the process by which we generated dipoles from poles, and so we find that a long thin solenoid acts like two poles (one positive, one negative) at its two ends. Finally, we find that circuits still act as dipole sheets when we examine the force exerted by one circuit on another. It is a bit inconvenient, however, to have to deal with surface dipole distributions that do not directly correspond to anything observable; at this point, therefore, some alternative formulations of the current-current force law prove expedient.

The modern preference is for a treatment that *begins* with the current-current forces, in a form that does not involve the dipole-sheet concept. We then proceed to current-magnet and magnet-magnet forces. In this development, the role previously played by the dipole defined as a limiting case of two poles is now played by something that we may still call a dipole, but that is defined as a limiting case of a small circuit: namely, let the area S shrink to zero under the constraint $I\boldsymbol{S}/c = \text{const} \equiv \boldsymbol{m}$. The pole concept is then never encountered, unless we choose to develop it from the properties of a long thin solenoid. This modern type of treatment has only one drawback, its greater mathematical complexity. In compensation, it has greater unity; and for that reason I will adopt it here. The unity is to some extent artificial, since an interpretation of electron spin as a current circuit is of dubious legitimacy; we must not take our unity too seriously.

2.2. Current-current forces. AMPERE's analysis of the experimental facts (see MASON and WEAVER [*17*], pp. 173—184) culminated in the following formula for the force $d\boldsymbol{F}_{12}$ exerted by a complete filamentary

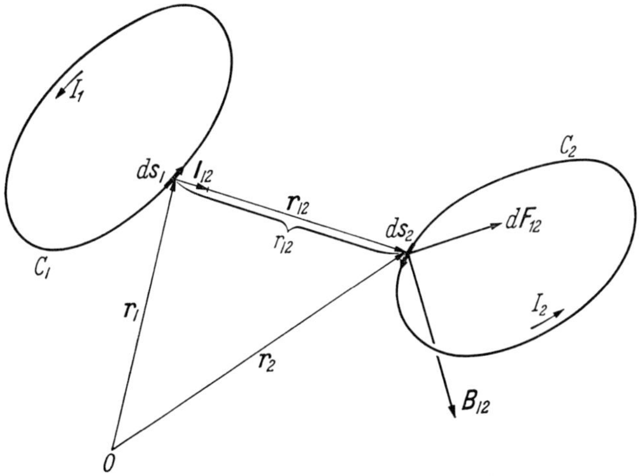

Fig. 3. Notation for Eq. (2.2), the force $d\boldsymbol{F}_{12}(=I_2\, d\boldsymbol{s}_2 \times \boldsymbol{B}_{12}/c)$ exerted by circuit C_1 on an element $I_2\, d\boldsymbol{s}_2$ of circuit C_2

circuit C_1, carrying current I_1, on an element $d\boldsymbol{s}_2$ of a filamentary circuit C_2, carrying current I_2 (see Fig. 3):

$$d\boldsymbol{F}_{12} = \frac{\gamma}{4\pi}\,\frac{I_2\, d\boldsymbol{s}_2}{c} \times \oint\limits_{C_1} \frac{I_1\, d\boldsymbol{s}_1}{c} \times \frac{\mathbf{1}_{12}}{r_{12}^2}\,. \tag{2.2}$$

This may be factored into

$$d\boldsymbol{F}_{12} = \frac{I_2\, d\boldsymbol{s}_2}{c} \times \boldsymbol{B}_{12}, \tag{2.3}$$

where

$$\boldsymbol{B}_{12} = \frac{\gamma}{4\pi} \oint\limits_{C_1} \frac{I_1\, d\boldsymbol{s}_1}{c} \times \frac{\mathbf{1}_{12}}{r_{12}^2}\,. \tag{2.4}$$

In Eq. (2.4) all reference to the presence of the element $d\boldsymbol{s}_2$ has disappeared; we may study the vector field $\boldsymbol{B}_{12}(\boldsymbol{r}_2)$ at an arbitrary point $\boldsymbol{r}_2$ of space, except on the singular line C_1. We find

$$\boldsymbol{B}_{12} = \nabla_2 \times \boldsymbol{A}_{12}, \tag{2.5}$$

with

$$\boldsymbol{A}_{12} = \frac{\gamma}{4\pi} \oint\limits_{C_1} \frac{I_1\, d\boldsymbol{s}_1}{c\, r_{12}}\,. \tag{2.6}$$

Thus $\boldsymbol{B}_{12}$, the *magnetic field intensity* (a term to be qualified later) of the circuit C_1, can be found by differentiation, with respect to the coordinates of the field point $\boldsymbol{r}_2$, of the *vector potential* $\boldsymbol{A}_{12}$ of the circuit.

The transition to a continuous volume or surface distribution of current is straightforward: we merely replace $I_1 d\boldsymbol{s}_1$ by $\boldsymbol{J}_1 d\tau_1$ or $\boldsymbol{K}_1 d S_1$, where $\boldsymbol{J}_1$ is the volume current density and $\boldsymbol{K}_1$ the surface current density (a vector tangent to the surface S_1). All these formulas apply strictly only to currents independent of time and satisfying the condition $\nabla \cdot \boldsymbol{J} = 0$ or one of its limiting forms.[1]

There is no difficulty about defining $\boldsymbol{A}$ or $\boldsymbol{B}$, or about carrying the operator ∇_2 in Eq. (2.5) past the integral sign in Eq. (2.6), at a point of the volume distribution; and from the integral definitions may be derived the differential relations

$$\nabla \cdot \boldsymbol{B} = 0, \qquad \nabla \times \boldsymbol{B} = \gamma \boldsymbol{J}/c. \tag{2.7}$$

The analogous boundary conditions at a surface occupied by a surface distribution are

$$\boldsymbol{n} \cdot \boldsymbol{B}^{+} - \boldsymbol{n} \cdot \boldsymbol{B}^{-} = 0, \qquad \boldsymbol{n} \times \boldsymbol{B}^{+} - \boldsymbol{n} \times \boldsymbol{B}^{-} = \gamma \boldsymbol{K}/c. \tag{2.8}$$

What physical significance, if any, to attach to the value of $\boldsymbol{B}$ at a point of the current distribution is a question that need not concern us; we shall suppose that the coils acting on our magnetic bodies are spatially separate from them and that eddy currents in the bodies are negligible. The practical reason for defining $\boldsymbol{B}$ (and $\boldsymbol{A}$) at internal points of the distributions is not that we wish to calculate mutual forces exerted between overlapping current distributions, but that one of the most powerful methods of calculating $\boldsymbol{B}$ is to solve the differential equations (2.7); this is often easier than evaluation of the integrals in the original definitions. For this purpose, however, we need one additional property: regularity at infinity. This means that for sources confined within a finite region, $|r^2 \boldsymbol{B}|$ and $|r \boldsymbol{A}|$ remain finite as the distance r to the field point increases without limit.

For a "small" circuit, *i.e.*, at distances r large in comparison with the circuit dimensions, the vector potential is approximately

$$\boldsymbol{A}_{12} = \frac{\gamma}{4\pi} \frac{\boldsymbol{m}_1 \times \boldsymbol{1}_{12}}{r_{12}^2}, \tag{2.9}$$

---

[1] A volume current density independent of time but with $\nabla \cdot \boldsymbol{J} \neq 0$ produces, by the continuity equation (conservation of charge), a time-rate of increase of charge density $\dot{\varrho} = -\nabla \cdot \boldsymbol{J}$; to calculate the resulting time-dependent fields requires use of MAXWELL's equations in their complete form. For the filamentary circuits considered in Eq. (2.2), the limiting form of $\nabla \cdot \boldsymbol{J} = 0$ is the constancy of the current I around a closed path C.

where (see Fig. 4) $\boldsymbol{r}_{12}$ is now drawn from a single, more or less centrally located, point of the circuit, and where

$$\boldsymbol{m}_1 \equiv \frac{I_1}{c}\, \boldsymbol{S}_1 = \frac{I_1}{c}\, S_1 \boldsymbol{n}_1 : \tag{2.10}$$

$\boldsymbol{S}_1$ is the vector area of the circuit C_1, *i.e.*, a vector whose i component is the projected area normal to the x_i axis; $S_1 = |\boldsymbol{S}_1|$, so that $\boldsymbol{n}_1$ is a unit vector. For a plane circuit, S_1 is the actual area and $\boldsymbol{n}_1$ is its positive unit normal. $\boldsymbol{m}_1$ is the *magnetic moment* of the circuit.

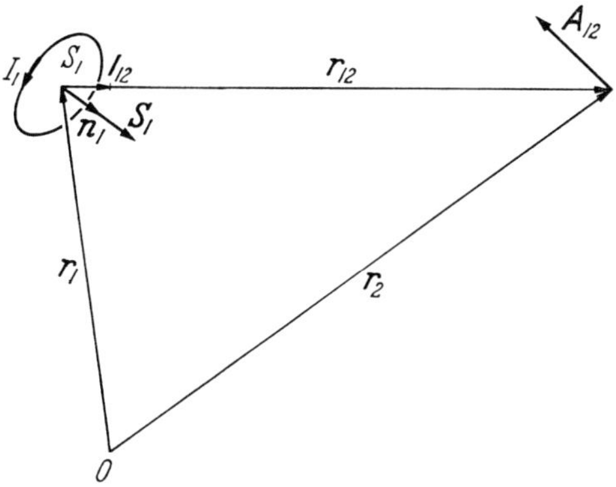

Fig. 4. Notation for Eq. (2.9), definition of the magnetic moment $\boldsymbol{m}_1 = I_1 \boldsymbol{S}_1/c$ of a circuit on the basis of its vector potential at distant points

We also find that the force and torque exerted on a small circuit C_2, with moment $\boldsymbol{m}_2$, by sources at a distance large in comparison with the dimensions of the circuit, are approximately

$$\left.\begin{aligned} \boldsymbol{F}_{12} &= \boldsymbol{m}_2 \cdot \nabla_2 \boldsymbol{B}_{12}, \\ \boldsymbol{L}_{12} &= \boldsymbol{r}_2 \times \boldsymbol{F}_{12} + \boldsymbol{m}_2 \times \boldsymbol{B}_{12}. \end{aligned}\right\} \tag{2.11}$$

In a uniform field, these reduce to zero force and a couple $\boldsymbol{m}_2 \times \boldsymbol{B}_{12}$, *i.e.*, a torque independent of the point about which it is computed.

Eqs. (2.9) and (2.11) become exact if we let the size of the circuit decrease to zero, its current meanwhile increasing without limit in such a way that the magnetic moment remains constant. Thus we arrive at the abstract concept of a (magnetic) *point dipole*. By taking the curl of $\boldsymbol{A}_{12}$, we obtain the field intensity of a dipole:

$$\boldsymbol{B}_{12} = \nabla_2 \times \boldsymbol{A}_{12} = \frac{\gamma}{4\pi} \, \frac{-\boldsymbol{m}_1 + 3\, \boldsymbol{m}_1 \cdot \mathbf{1}_{12} \mathbf{1}_{12}}{r_{12}^3}\,. \tag{2.12}$$

If we define a dipole on the basis of poles, in the manner sketched in § 2.1, we get for its field intensity

$$\boldsymbol{B}_{12} = -\nabla_2 \varphi_{12}, \tag{2.13}$$

where the *scalar potential* φ is given by

$$\varphi_{12} = \frac{\gamma}{4\pi} \frac{\boldsymbol{m}_1 \cdot \mathbf{1}_{12}}{r_{12}^2}. \tag{2.14}$$

Performance of the differentiation leads again to the third member of Eq. (2.12). Thus the two definitions of a magnetic dipole lead to the same $\boldsymbol{B}$ at any point except the position of the dipole, and to the same singularity there. But at any stage short of the limit, the field intensity between the two poles $\pm p$ points opposite to the direction of $\boldsymbol{m}$, whereas

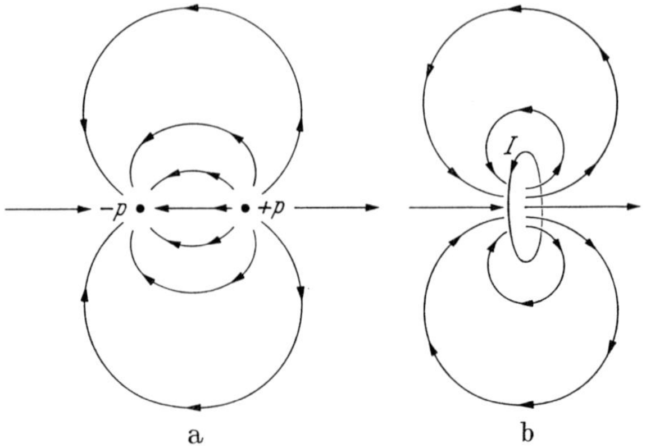

Fig. 5 (a) and (b). Comparison of the field between the poles of a doublet with the field inside the current path in a small plane circuit

the field intensity at the center of a small plane circular circuit points along the direction of $\boldsymbol{m}$ (see Fig. 5). If, therefore, we replace a small but finite circuit or a small but finite pair of poles by a dipole, we have negligible error at large distances, but we have thrown away all information about the size and even the sign of the field at distances comparable with the dimensions of the circuit or doublet. If we now go over to a continuous volume distribution of dipoles, we shall get rid of the singularity and be able to calculate finite fields at points of the distribution; but we must not expect these calculated fields to have any physical significance for an actual distribution of small circuits or of small doublets within a body. We cannot, by spreading out a point singularity into a finite smear, recover information that we lost in the limiting process that produced the singularity. Loss of information is an irreversible process (FISHER [*1*], SHANNON [*1*], BRILLOUIN [*1*]).

2.3. Magnetized bodies. Small magnetized bodies, as regards the mutual forces between two of them or between one of them and a current circuit, behave like small circuits. This concise statement summarizes a number of different experimentally observed phenomena. To extend our theory so as to cover such phenomena, we need only to attribute to each small body a magnetic moment $\boldsymbol{m}$, chosen to fit the experimental facts in any particular experiment. To calculate the forces exerted *by* a small magnetized body, we calculate its field intensity by Eq. (2.12), then use Eq. (2.3) to find the forces exerted on a distant circuit element, Eqs. (2.11) to find the forces exerted on another small body. To calculate the forces exerted *on* a small magnetized body, we calculate the field intensities of circuits by Eq. (2.4) or its later equivalents and the field intensities of other small bodies by Eq. (2.12), then use Eq. (2.11) to find the resulting forces.

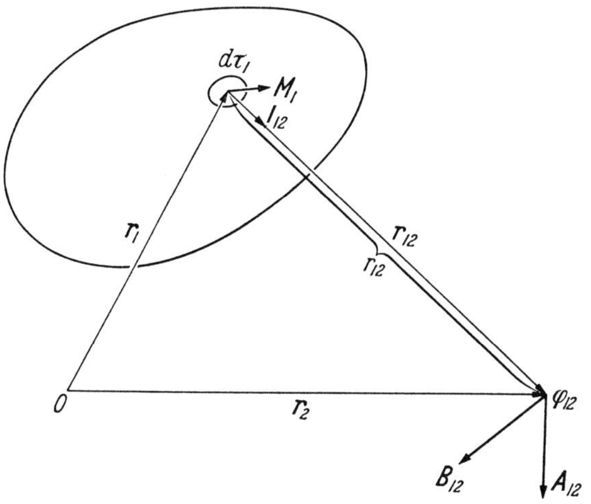

Fig. 6. Notation for Eq. (2.15): the external field intensity $\boldsymbol{B}_{12}$, vector potential $\boldsymbol{A}_{12}$, and scalar potential φ_{12} of a magnetized body (with magnetization $\boldsymbol{M}_1$) at an external point $\boldsymbol{r}_2$

"Small" means that the dimensions of the magnetized body are small in comparison with its distance to the circuits and other bodies considered. When this is not so, the body must be represented by a *distribution* of moments. Thus we are led to the concept of *magnetic moment per unit volume*, or magnetization, $\boldsymbol{M}$, a vector field defined within a physical body and such that the magnetic moment assigned to volume element $d\tau$ is $\boldsymbol{M}\,d\tau$.

The magnetic field due to a finite magnetized body, at a point outside it (and distant from it by a large number of lattice spacings or intermolecular distances, so that the approximations of continuum theory are justified), is then, by Eqs. (2.12) to (2.14) and (2.9),

$$\boldsymbol{B}_{12} = \frac{\gamma}{4\pi}\int \frac{-\boldsymbol{M}_1 + 3\boldsymbol{M}_1 \cdot \boldsymbol{1}_{12}\boldsymbol{1}_{12}}{r_{12}^3}\,d\tau_1 = \nabla_2 \times \boldsymbol{A}_{12} = -\nabla_2\varphi_{12} \qquad (2.15)$$

(see Fig. 6), where

$$\boldsymbol{A}_{12} = \frac{\gamma}{4\pi}\int \frac{\boldsymbol{M}_1 \times \boldsymbol{1}_{12}}{r_{12}^2}\,d\tau_1 \qquad (2.16)$$

and

$$\varphi_{12} = \frac{\gamma}{4\pi} \int \frac{\boldsymbol{M}_1 \cdot \boldsymbol{1}_{12}}{r_{12}^2} \, d\tau_1. \tag{2.17}$$

At external points, all three expressions in Eq. (2.15) are equivalent. Furthermore, by noting that ($r \equiv r_{12}$)

$$\boldsymbol{1}_{12}/r^2 = \nabla_1(1/r) \tag{2.18}$$

and using the vector identities

$$\nabla_1 \times (\boldsymbol{M}_1/r) = (\nabla_1 \times \boldsymbol{M}_1)/r - \boldsymbol{M}_1 \times \nabla_1(1/r), \tag{2.19}$$

$$\nabla_1 \cdot (\boldsymbol{M}_1/r) = (\nabla_1 \cdot \boldsymbol{M}_1)/r + \boldsymbol{M}_1 \cdot \nabla_1(1/r), \tag{2.20}$$

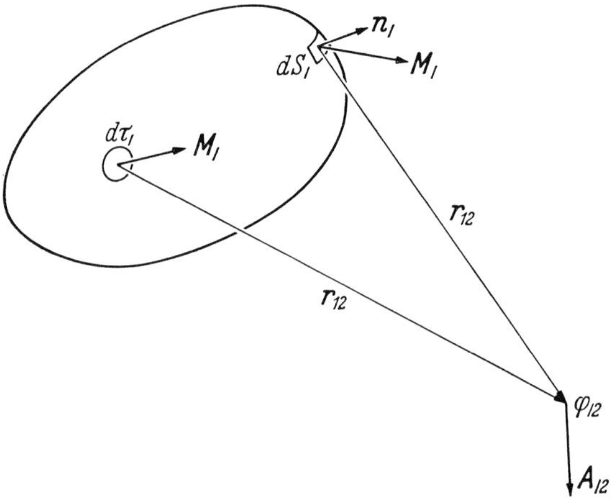

Fig. 7. Notation for Eqs. (2.21), (2.22): alternative formulas for the vector potential $\boldsymbol{A}_{12}$ and scalar potential φ_{12} of a magnetized body (with magnetization $\boldsymbol{M}_1$) at an external point $\boldsymbol{r}_2$

and appropriate volume-to-surface integral transformations, we can put $\boldsymbol{A}_{12}$ and φ_{12} into the form (see Fig. 7)

$$\boldsymbol{A}_{12} = \frac{\gamma}{4\pi} \left\{ \int \frac{\nabla_1 \times \boldsymbol{M}_1}{r} \, d\tau_1 + \int \frac{(-\boldsymbol{n}_1 \times \boldsymbol{M}_1)}{r} \, dS_1 \right\}, \tag{2.21}$$

$$\varphi_{12} = \frac{\gamma}{4\pi} \left\{ \int \frac{(-\nabla_1 \cdot \boldsymbol{M}_1)}{r} \, d\tau_1 + \int \frac{\boldsymbol{n}_1 \cdot \boldsymbol{M}_1}{r} \, dS_1 \right\}. \tag{2.22}$$

Eq. (2.21) shows that $\boldsymbol{A}_{12}$ (and hence $\boldsymbol{B}_{12}$) can be calculated as if it were produced by volume and surface currents of densities $c\nabla_1 \times \boldsymbol{M}_1$ and $-c\boldsymbol{n}_1 \times \boldsymbol{M}_1$ respectively. Eq. (2.22) shows that φ_{12} (and hence $\boldsymbol{B}_{12}$) can be calculated as if it were produced by volume and surface pole densities [cf. Eq. (2.1); the field intensity of p_1 is defined as $\boldsymbol{F}_{12}/p_2$] $-\nabla_1 \cdot \boldsymbol{M}_1$ and $\boldsymbol{n}_1 \cdot \boldsymbol{M}_1$ respectively. The two formulas for $\boldsymbol{B}_{12}$ so obtained are equivalent to each other and to the first Eq (2.14), provided the field point lies outside the body 1.

2.4. The field vectors $\boldsymbol{B}$ and $\boldsymbol{H}$. In §2.3, it was supposed that the field point $\boldsymbol{r}_2$ lay outside the body whose magnetization $\boldsymbol{M}_1$ was responsible for the field. We now investigate what meaning, if any, the formulas have at internal points. Here the integrals over the body volume τ_1 must be defined (see Fig. 8) as limiting values, as δ shrinks to zero, of the integrals over the volume $\tau_1 - \delta$ inside the body but outside a small closed surface δ surrounding the point $r_{12}=0$ (LEATHEM [*1*]). The integral in the second member of Eq. (2.15) is semiconvergent: its value depends on the shape of δ. The integrals (2.16) and (2.17) converge but may not be differentiated under the integral sign; they may, however, still be transformed to the forms (2.21), (2.22). The integrals

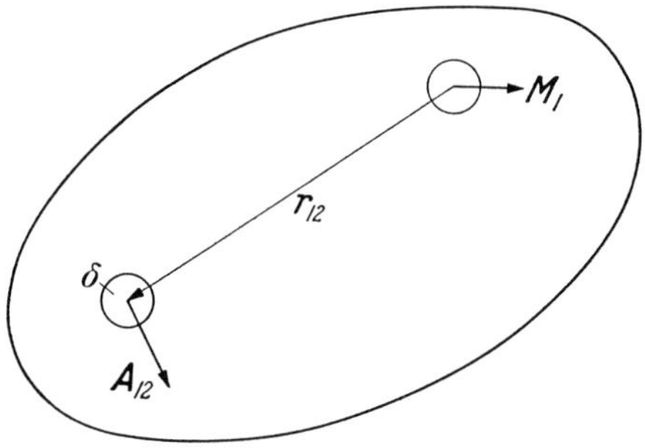

Fig. 8. Definition of $\boldsymbol{A}_{12}$, φ_{12}, etc. at a point $\boldsymbol{r}_2$ inside the body whose magnetization $\boldsymbol{M}_1$ is the source of the field. Since the integrand becomes infinite when $r_{12} \to 0$, a small region δ about the point $\boldsymbol{r}_2$ must be initially excluded from the integration; then δ is shrunk to zero. In order that the result may have a unique meaning, it must not merely be finite: it must also be independent of the shape, during the limiting process, of δ

(2.21) and (2.22) converge and may be differentiated under the integral sign, but the two do not give the same result. We write ($\nabla \equiv \nabla_2$)

$$\boldsymbol{B}_{12} \equiv \nabla \times \boldsymbol{A}_{12}, \qquad \boldsymbol{H}_{12} \equiv -\nabla \varphi_{12}; \tag{2.23}$$

then it may be shown that

$$\boldsymbol{B}_{12} - \boldsymbol{H}_{12} = \gamma \boldsymbol{M}_2. \tag{2.24}$$

Thus the small-circuit and pole-doublet representations of magnetized bodies lead to the same calculated "field intensity" at external points ($\boldsymbol{M}_2 = 0$), but not at internal. This should occasion no surprise in view of the remarks at the end of § 2.1.

We shall not use the term "field intensity" at internal points of the body producing the field under consideration. If we add to the field vectors $\boldsymbol{B}_{12}$ and $\boldsymbol{H}_{12}$, just defined, the similarly defined field vectors due to other magnetized bodies, and also the field intensity of the magnetizing currents, we get two field vectors $\boldsymbol{B}$ and $\boldsymbol{H}$ that are identical outside magnetized bodies and that are related by the equation

$$\boldsymbol{B} - \boldsymbol{H} = \gamma \boldsymbol{M} \tag{2.25}$$

(we may now drop the subscripts). We shall call $\boldsymbol{B}$ the (*magnetic*) *induction* and $\boldsymbol{H}$ the *magnetizing force*; the names are bad ones, chosen arbitrarily from a number of commonly used names that include no good ones. *We shall attribute no physical significance whatsoever to* $\boldsymbol{B}$ *and* $\boldsymbol{H}$ *at internal points*; a sufficient justification for defining them there is that when so defined, they satisfy differential equations and boundary conditions that can be solved as a means of calculating them at *all* points, including the external points at which they do have physical significance.

The differential equations and boundary conditions may be stated most simply as follows: in a region where there are no "conduction" currents $\boldsymbol{J}_c$ and $\boldsymbol{K}_c$, $\boldsymbol{B}$ is solenoidal and $\boldsymbol{H}$ is irrotational. "Solenoidal" implies not only $\nabla\cdot\boldsymbol{B}=0$ but, at a surface of discontinuity, $\boldsymbol{n}\cdot\boldsymbol{B}^+-\boldsymbol{n}\cdot\boldsymbol{B}^-=0$; "irrotational" implies not only $\nabla\times\boldsymbol{H}=0$ but, at a surface of discontinuity, $\boldsymbol{n}\times\boldsymbol{H}^+-\boldsymbol{n}\times\boldsymbol{H}^-=0$. In a region where there are conduction currents as well as, perhaps, magnetization,

$$\nabla\cdot\boldsymbol{B}=0,\qquad \nabla\times\boldsymbol{H}=(\gamma/c)\,\boldsymbol{J}_c; \tag{2.26}$$

these, by Eq. (2.25), are equivalent to

$$\nabla\cdot\boldsymbol{H}=-\gamma\nabla\cdot\boldsymbol{M},\qquad \nabla\times\boldsymbol{B}=(\gamma/c)\,\boldsymbol{J}_c+\gamma\nabla\times\boldsymbol{M}. \tag{2.27}$$

The second Eq. (2.27) may be written

$$\nabla\times\boldsymbol{B}=\gamma\boldsymbol{J}, \tag{2.28}$$

where

$$\boldsymbol{J}=\boldsymbol{J}_c+c\nabla\times\boldsymbol{M} \tag{2.29}$$

is the "total" (conduction plus magnetization) volume current density. On the surface of a magnetized body where there are surface conduction currents,

$$\boldsymbol{n}\cdot(\boldsymbol{B}^+-\boldsymbol{B}^-)=0,\qquad \boldsymbol{n}\times(\boldsymbol{H}^+-\boldsymbol{H}^-)=\gamma\boldsymbol{K}_c; \tag{2.30}$$

these, by Eq. (2.25), are equivalent to

$$\boldsymbol{n}\cdot(\boldsymbol{H}^+-\boldsymbol{H}^-)=\gamma\boldsymbol{n}\cdot\boldsymbol{M},\qquad \boldsymbol{n}\times(\boldsymbol{B}^+-\boldsymbol{B}^-)=(\gamma/c)\,\boldsymbol{K}_c-\gamma\boldsymbol{n}\times\boldsymbol{M} \tag{2.31}$$

if the positive side of the surface is *outside* with respect to the magnetized body. The second Eq. (2.31) may be written

$$\boldsymbol{n}\times(\boldsymbol{B}^+-\boldsymbol{B}^-)=(\gamma/c)\,\boldsymbol{K}, \tag{2.32}$$

where

$$\boldsymbol{K}=\boldsymbol{K}_c-c\boldsymbol{n}\times\boldsymbol{M} \tag{2.33}$$

is the "total" surface current density.

It is a question of convenience only whether to use $\boldsymbol{B}$, whose sources are conduction currents $\boldsymbol{J}_c$, $\boldsymbol{K}_c$ and magnetization "currents" $c\nabla\times\boldsymbol{M}$, $-c\boldsymbol{n}\times\boldsymbol{M}$, or to use $\boldsymbol{H}$, whose sources are the same conduction currents together with magnetization "charges" or poles, of volume density $-\nabla \cdot \boldsymbol{M}$ and surface density $\boldsymbol{n} \cdot \boldsymbol{M}$.

Altho the differential equations will be useful to us, in some cases it will be more useful to express $\boldsymbol{B}_{12}$ and $\boldsymbol{H}_{12}$ directly in terms of their sources, and without use of $\boldsymbol{A}_{12}$ and φ_{12}. From the previous discussion, it is clear that the direct formulas are

$$\boldsymbol{B}_{12} = \frac{\gamma}{4\pi}\left\{\int\limits_{S_1} (-\boldsymbol{n}_1 \times \boldsymbol{M}_1) \times \frac{\boldsymbol{1}_{12}}{r_{12}^2}\, dS_1 + \int\limits_{\tau_1} (\nabla_1 \times \boldsymbol{M}_1) \times \frac{\boldsymbol{1}_{12}}{r_{12}^2}\, d\tau_1\right\}, \tag{2.34}$$

$$\boldsymbol{H}_{12} = \frac{\gamma}{4\pi}\left\{\int\limits_{S_1} (\boldsymbol{n}_1 \cdot \boldsymbol{M}_1) \frac{\boldsymbol{1}_{12}}{r_{12}^2}\, dS_1 + \int\limits_{\tau_1} (-\nabla_1 \cdot \boldsymbol{M}_1) \frac{\boldsymbol{1}_{12}}{r_{12}^2}\, d\tau_1\right\}. \tag{2.35}$$

2.5. The force and torque on a complete body. The fact that the field vectors $\boldsymbol{B}$ and $\boldsymbol{H}$ can be incorporated into practical working formulas, without attributing physical significance to them at internal points, will now be illustrated by consideration of the magnetic force exerted on a magnetized and conduction-current-carrying body by all other currents and magnetized bodies. It is assumed that the body under consideration is separated from the other bodies by a region free from conduction currents and magnetization (see Fig. 9).

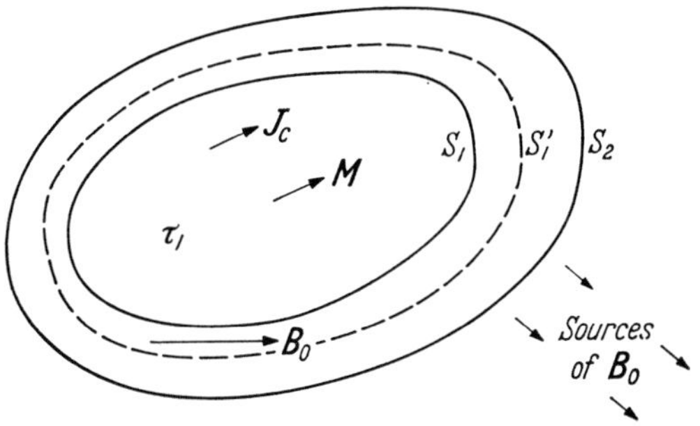

Fig. 9. Notation for Eqs. (2.36) ff.: calculation of the total force $\boldsymbol{F}$ and torque $\boldsymbol{L}$ exerted on a magnetized and current-carrying body 1 (occupying the region τ_1 bounded by surface S_1) by sources outside a surrounding surface S_2, whose magnetic field intensity is $\boldsymbol{B}_0$. The space between S_1 and S_2 contains no sources of magnetic field, and S_1' is in this region.

Let $\boldsymbol{B}_0$ be the field intensity of the other currents and magnetized bodies; more briefly, the "applied field" intensity acting on the body 1 under consideration. Within and on some surface S_1' surrounding this body, there are no sources of $\boldsymbol{B}_0$. From the basic force laws and definitions, the force and torque in question are

$$\boldsymbol{F} = \frac{1}{c}\int\limits_{\tau_1} \boldsymbol{J} \times \boldsymbol{B}_0\, d\tau, \qquad \boldsymbol{L} = \frac{1}{c}\int\limits_{\tau_1} \boldsymbol{r} \times (\boldsymbol{J} \times \boldsymbol{B}_0)\, d\tau, \tag{2.36}$$

where $\boldsymbol{J}$ is the *total* current density, $\boldsymbol{J}_c + c\nabla\times\boldsymbol{M}$. We shall regard the surface current densities ($\boldsymbol{K}_c$ and $-\boldsymbol{n}\times\boldsymbol{M}$) as limiting cases of volume distributions, an often convenient method of simplifying the mathematics. We now write $\boldsymbol{B}_0 = \boldsymbol{B} - \boldsymbol{B}_1$, where $\boldsymbol{B}_1$ is the contribution to $\boldsymbol{B}$ from body 1. Since $\boldsymbol{J} = (c/\gamma)\nabla\times\boldsymbol{B} = (c/\gamma)\nabla\times\boldsymbol{B}_1$, we have

$$\left.\begin{aligned} \boldsymbol{F}_1 &\equiv \frac{1}{c}\int\limits_{\tau_1'} \boldsymbol{J}\times\boldsymbol{B}_1\, d\tau = \frac{1}{\gamma}\int\limits_{\tau_1'} (\nabla\times\boldsymbol{B}_1)\times\boldsymbol{B}_1\, d\tau \\ &= \frac{1}{\gamma}\int\limits_{\tau_1'} \left[\boldsymbol{B}_1\cdot\nabla\boldsymbol{B}_1 - \frac{1}{2}\nabla(\boldsymbol{B}_1^2)\right] d\tau; \end{aligned}\right\} \tag{2.37}$$

we are now integrating over the volume τ_1' inside S_1'. The x_i component of the first integrand is

$$B_{1k}B_{1i,k} = (B_{1k}B_{1i})_{,k} - B_{1k,k}B_{1i} = (B_{1k}B_{1i})_{,k}, \quad \text{since} \quad B_{1k,k} = \nabla\cdot\boldsymbol{B}_1 = 0.$$

Hence

$$\left.\begin{aligned} F_{1i} &= \frac{1}{\gamma}\int\limits_{\tau_1'} \left[(B_{1k}B_{1i})_{,k} - \frac{1}{2}(\boldsymbol{B}_1^2)_{,i}\right] d\tau \\ &= \frac{1}{\gamma}\int\limits_{S_1'} \left[B_{1k}B_{1i}n_k - \frac{1}{2}\boldsymbol{B}_1^2 n_i\right] dS. \end{aligned}\right\} \tag{2.38}$$

Eq. (2.38) in vector form is

$$\boldsymbol{F}_1 = \frac{1}{\gamma}\int\limits_{S_1'} \left[\boldsymbol{n}\cdot\boldsymbol{B}_1\boldsymbol{B}_1 - \frac{1}{2}\boldsymbol{n}\boldsymbol{B}_1^2\right] dS. \tag{2.39}$$

The same steps, with $\boldsymbol{B}$ replacing $\boldsymbol{B}_1$, give

$$\left.\begin{aligned} \boldsymbol{F} - \boldsymbol{F}_1 &= \frac{1}{c}\int\limits_{\tau_1'} \boldsymbol{J}\times\boldsymbol{B}\, d\tau \\ &= \frac{1}{\gamma}\int\limits_{S_1'} \left[\boldsymbol{n}\cdot\boldsymbol{B}\boldsymbol{B} - \frac{1}{2}\boldsymbol{n}\boldsymbol{B}^2\right] dS. \end{aligned}\right\} \tag{2.40}$$

Furthermore, the same steps applied to the volume between S_1' and a second surface S_1'' surrounding it, so that between the two surfaces $\nabla\times\boldsymbol{B}_1 = 0$, show that in Eq. (2.39) [but not in Eq. (2.40)!] the surface S_1' may be replaced by *any* surface surrounding the body 1. Let this alternative surface now expand to infinity. Since $\boldsymbol{B}_1$ is the field intensity of sources within the finite region τ_1, it decays at infinity at least as fast as $1/r^2$ (actually, since there are no isolated magnetic poles, at least as fast as $1/r^3$); $dS = r^2 d\omega$ (where $d\omega$ is an element of solid angle) increases as r^2; therefore the integral decreases at least as fast as $r^{-2}\cdot r^{-2}\cdot r^2 = 1/r^2$. Thus $\boldsymbol{F}_1 = 0$, and from Eq. (2.40)

$$\boldsymbol{F} = \frac{1}{\gamma}\int\limits_{S_1'} \left[\boldsymbol{n}\cdot\boldsymbol{B}\boldsymbol{B} - \frac{1}{2}\boldsymbol{n}\boldsymbol{B}^2\right] dS. \tag{2.41}$$

A similar argument applied to the torque $\boldsymbol{L}$ gives for it a formula that can be obtained from Eq. (2.41) by inserting $\boldsymbol{r}$ before the quantity in brackets. In either of these formulas, $\boldsymbol{B}$ may be replaced by $\boldsymbol{H}$, since $\boldsymbol{M}=0$ on S_1'.

Formula (2.41) is the famous "Maxwell stress" formula. It enables us to calculate the total force and torque on a body by an integration over a surface surrounding it. The force in question is that exerted by currents and magnetization outside the surface; if we assume that the body exerts no magnetic force or torque on itself, this is the total magnetic force. Eq. (2.41) says nothing about the magnetic force exerted on *part* of a body; nor does the derivation justify our attributing any meaning to an analogous integral over a surface that lies partly or wholly within the body 1.

MAXWELL [*1*] (Vol. I, pp. 155ff.) derived the electrostatic analog of Eq. (2.41) by an argument analogous to that given here. In the magnetic case, however (Vol. II, pp. 276ff.), he used a different argument. He started with the following formula for the magnetic force on "an element of a body" $d\tau$ that both carries conduction current and is magnetized:

$$\text{“}dF_i=[\boldsymbol{M}\cdot\partial\boldsymbol{H}/\partial x_i+(\boldsymbol{J}_{\mathrm{c}}\times\boldsymbol{B})_i]\,d\tau.\text{”} \tag{2.42}$$

The force on the elements within a region bounded by an arbitrary surface S can then be expressed in the form

$$\left.\begin{aligned}\text{“}\boldsymbol{F}&=\frac{1}{\gamma}\int_S\left[\boldsymbol{n}\cdot\boldsymbol{B}\boldsymbol{H}-\frac{1}{2}\boldsymbol{n}H^2\right]dS\\&=\frac{1}{\gamma}\int_S\left[\boldsymbol{n}\cdot\boldsymbol{H}\boldsymbol{H}+\gamma\boldsymbol{n}\cdot\boldsymbol{M}\boldsymbol{H}-\frac{1}{2}\boldsymbol{n}H^2\right]dS.\text{”}\end{aligned}\right\} \tag{2.43}$$

When S surrounds the whole body, Eq. (2.43) reduces to Eq. (2.41). It is clear, however, that Eq. (2.43) goes further than Eq. (2.41), since it purports to evaluate the force on the matter in an arbitrary region. Its meaning and validity are dependent on the meaning and validity of Eq. (2.42). The terms in this equation, however, had been derived previously (MAXWELL, [*1*] §§ 389, 489, 602—603) only for the case in which the field involved was that of external sources.[1]

---

[1] These previous discussions exploited the irrotational property of the field in the case of the first term and its solenoidal property in the case of the second; this may have been MAXWELL's reason for writing $\boldsymbol{H}$ in the first and $\boldsymbol{B}$ in the second. His posthumous editors ([ ] = Niven, { } = Thomson; see prefaces, Vol. I) were disturbed by the fact that in the new application, $\boldsymbol{H}$ was no longer irrotational (p. 285, Appendix II). THOMSON therefore started with the equation "$dF_i=[\boldsymbol{M}\cdot\nabla H_i+(\boldsymbol{J}\times\boldsymbol{H})_i]\,d\tau$" and transformed it to the form (2.42).

Toupin [1] based his theory on the electrostatic analog of Eq. (2.43) for an arbitrary region.[1] *We shall not use this equation,* or its ancestor Eq. (2.42) — which, with $\boldsymbol{J}_c = 0$ and therefore $\nabla \times \boldsymbol{H} = 0$, so that $\partial \boldsymbol{H}/\partial x_i = \nabla H_i$, is the force formula postulated by Tiersten [1]. Our reason for rejecting it is that it is a *superfluous postulate,* in the sense discussed in § 1.4. This will be shown in detail in Chapter II. Briefly, the formula arbitrarily selects a part of an observable force, not itself separately observable, and proclaims this to be the "magnetic" force and the rest to be the "mechanical". Its content is therefore not physical but semantic.

We shall, instead, use formulas equivalent to Eqs. (2.36) for the magnetic force and torque on *a whole body* (see § 5.3). Our introduction of $\boldsymbol{B}$ or $\boldsymbol{H}$ will be based, as here, on the formulas $\boldsymbol{B}_0 = \boldsymbol{H}_0 = \boldsymbol{B} - \boldsymbol{B}_1 = \boldsymbol{H} - \boldsymbol{H}_1$, and not on any supposed physical meaning of $\boldsymbol{B}$ or $\boldsymbol{H}$. The transformations that prove convenient, however, will be somewhat different ones from those just used.

2.6. Energy relations for a rigid magnetic body. To calculate the work done in magnetizing a body, we must suppose that the currents in the magnetizing coils change slowly enough not to invalidate the magnetostatic formulas of the previous sections; this will be true if only a negligibly small change occurs during the time interval required for propagation of an electromagnetic effect, at speed c, from any point of the system to any other. We suppose in addition that the magnetizing coil, which carries current I and produces the applied field $\boldsymbol{H}_0$, is rigid and stationary. Then the time rate of work by the battery[2], by virtue of the presence of the magnetized body, is $-I\mathscr{E}_c$, where $\mathscr{E}_c$ is the induced electromotive force

$$\mathscr{E}_c = -\frac{1}{c}\, d\Phi_m/dt \tag{2.44}$$

produced by change of magnetization of the body. Here (see Fig. 10) Φ_m is the flux of the body's induction $\boldsymbol{B}_1$ thru the coil C_2,

$$\Phi_m = \int_{S_2} \boldsymbol{n}_2 \cdot \boldsymbol{B}_{12}\, dS_2 = \oint_{C_2} \boldsymbol{A}_{12} \cdot d\boldsymbol{s}_2; \tag{2.45}$$

S_2 is a surface bounded by C_2, and $\boldsymbol{A}_{12}$ is the vector potential of the magnetized body. The last step follows by setting $\boldsymbol{B}_{12} = \nabla_2 \times \boldsymbol{A}_{12}$ and using Stokes's theorem. On putting these relations together, and writing δW for the work in time interval δt (the symbol δ is a reminder

[1] This is my own inference about where Toupin found the formula. He himself introduced it with the remark, "It is known from Maxwell's work that ..."; his only reference was the two volumes of Maxwell [1].

[2] The term "work by the battery" is short for "work by the electromotive force of the battery"; *cf.* the discussion of work in § 4.1.

that δW need not be a perfect differential), we have

$$\frac{\delta W}{\delta t}=\frac{I}{c}\frac{d\Phi_{\mathrm{m}}}{dt}=\frac{I}{c}\frac{d}{dt}\oint_{C_2} \boldsymbol{A}_{12}\cdot d\boldsymbol{s}_2. \tag{2.46}$$

In case the body is rigid, the calculation is simple:

$$\frac{\delta W}{\delta t}=\frac{I}{c}\oint_{C_2}\frac{\partial \boldsymbol{A}_{12}}{\partial t}\cdot d\boldsymbol{s}_2=\int_{\tau_1}\boldsymbol{H}_0\cdot\frac{\partial \mathbf{M}_1}{\partial t}\,d\tau_1, \tag{2.47}$$

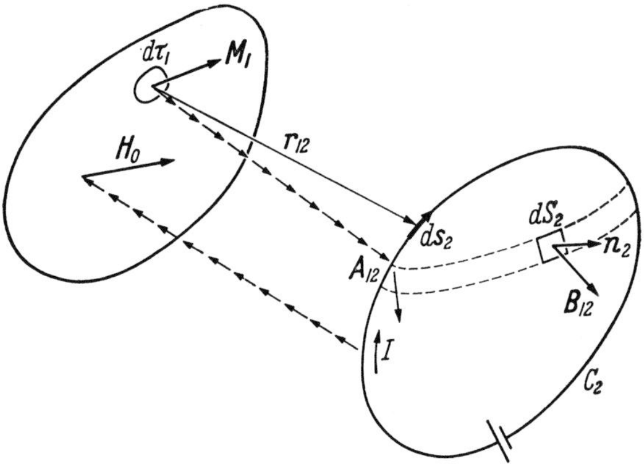

Fig. 10. Notation for Eqs. (2.45) ff. The coil C_2 with current I produces applied field intensity $\boldsymbol{H}_0$ in the body 1, whose magnetization is $\boldsymbol{M}_1$. The body produces vector potential $\boldsymbol{A}_{12}$ and magnetic induction $\boldsymbol{B}_{12}$ in the region occupied by the coil

by use of Eq. (2.16) for $\boldsymbol{A}_{12}$ and of Eq. (2.4) for $\boldsymbol{H}_0$ (with appropriate changes of the subscripts and notation). Hence (we now drop the subscript 1)

$$\delta W=\int_{\tau}\boldsymbol{H}_0\cdot\delta\boldsymbol{M}\,d\tau. \tag{2.48}$$

As in § 2.5, we may set $\boldsymbol{H}_0(=\boldsymbol{B}_0)=\boldsymbol{B}-\boldsymbol{B}_1=\boldsymbol{H}-\boldsymbol{H}_1$, where $\boldsymbol{B}_1(\boldsymbol{H}_1)$ is the part of $\boldsymbol{B}(\boldsymbol{H})$ that is due to the magnetized body. By well-known transformation theorems of magnetostatics (Brown [10], Chap. 3),

$$\left.\begin{aligned}-\int\boldsymbol{H}_1\cdot\delta\boldsymbol{M}\,d\tau&=-\int\boldsymbol{M}\cdot\delta\boldsymbol{H}_1\,d\tau\\&=-\tfrac{1}{2}\int(\boldsymbol{H}_1\cdot\delta\boldsymbol{M}+\boldsymbol{M}\cdot\delta\boldsymbol{H}_1)\,d\tau=\delta W_{\mathrm{m}},\end{aligned}\right\} \tag{2.49}$$

where

$$W_{\mathrm{m}}\equiv-\frac{1}{2}\int\boldsymbol{H}_1\cdot\boldsymbol{M}\,d\tau=\frac{1}{2\gamma}\int\boldsymbol{H}_1^2\,d\tau. \tag{2.50}$$

Similarly

$$-\int\boldsymbol{B}_1\cdot\delta\boldsymbol{M}\,d\tau=\delta W_{\mathrm{m}}', \tag{2.51}$$

where

$$W_{\mathrm{m}}'=-\frac{1}{2}\int\boldsymbol{B}_1\cdot\boldsymbol{M}\,d\tau=-\frac{1}{2\gamma}\int\boldsymbol{B}_1^2\,d\tau. \tag{2.52}$$

The integrals extend over all of space where the integrands differ from zero. Thus from Eq. (2.48) we find

$$\left.\begin{aligned}\delta W &= \delta W_{\mathrm{m}} + \int \boldsymbol{H} \cdot \delta \boldsymbol{M}\, d\tau\\ &= \delta W_{\mathrm{m}}' + \int \boldsymbol{B} \cdot \delta \boldsymbol{M}\, d\tau.\end{aligned}\right\} \tag{2.53}$$

We may define either W_{m} or W_{m}' as the "magnetostatic self-energy" of the body. The two are not equivalent, and neither is identical with the dipole-dipole energy calculated from a microscopic model (BROWN [*10*], pp. 101—102). Either differs from it, however, by the volume integral of a quantity that depends only on local conditions and may therefore be regarded as an energy (or free-energy) density. The difference

$$\left.\begin{aligned}W_{\mathrm{m}}' - W_{\mathrm{m}} &= -\tfrac{1}{2}\int (\boldsymbol{B}_1 - \boldsymbol{H}_1) \cdot \boldsymbol{M}\, d\tau\\ &= -\tfrac{1}{2}\gamma \int \boldsymbol{M}^2\, d\tau\end{aligned}\right\} \tag{2.54}$$

clearly has this property. Thus subtraction of either perfect differential, δW_{m} or $\delta W_{\mathrm{m}}'$, from δW removes from consideration the troublesome part of the dipole-dipole energy, the long-range part that depends on the shape of the specimen. This can be illustrated by the simple case of an ellipsoid uniformly magnetized along a principal axis. Here

$$\boldsymbol{H}_1 = -\gamma D \boldsymbol{M}, \qquad \boldsymbol{B}_1 = \gamma (1 - D) \boldsymbol{M}, \tag{2.55}$$

where D is the "demagnetizing factor" for that axis (STONER [*1*]). Therefore, if V is the specimen volume,

$$W_{\mathrm{m}} = \tfrac{1}{2}\gamma D M^2 V, \qquad W_{\mathrm{m}}' = -\tfrac{1}{2}\gamma (1 - D) M^2 V. \tag{2.56}$$

The term $\frac{1}{2}\gamma D M^2 V$ in either of these expressions contains the shape-dependent part of the dipole-dipole energy.

Having removed the shape-dependent part in this way, we may expect to relate the remaining part of δW, $\int \boldsymbol{H} \cdot \delta \boldsymbol{M}\, d\tau$ or $\int \boldsymbol{B} \cdot \delta \boldsymbol{M}\, d\tau$, to energy either lost in irreversible processes or stored in local form (expressible as the integral of a volume density of free energy) — unless, of course, the system contains other long-range forces, such as those between electric charges on its surface. In a complete cycle, $\int \oint \boldsymbol{H} \cdot \delta \boldsymbol{M}\, d\tau = \int \oint \boldsymbol{B} \cdot \delta \boldsymbol{M}\, d\tau$ measures the energy lost by hysteresis and other mechanisms.

We have omitted the work done by the battery against the electromotive force induced by the coil's own flux change. This is the change in the self-energy $\frac{1}{2} L_0 I^2$ of the coil, where L_0 is its self-inductance in the absence of the specimen. As long as the current I is directly controlled, this term in the work plays no role in the behavior of the specimen.

For a deformable body, the work and energy expressions are more complicated. They will be derived in § 6.1.

2.7. Ferromagnetic materials. From the phenomenological point of view, the defining characteristic of a ferromagnetic solid is that when examined on a sufficiently small scale (still much larger than the lattice spacing, so that continuous concepts may still be used), it has a magnetization $\boldsymbol{M}$ whose magnitude $M_s(T)$ is, to a good approximation, a function of temperature only; its direction, however, may vary with position. In this section, as in the preceding one, we shall assume that the body is rigid; the modifications necessitated by magnetostriction will be treated in later chapters.

The *spontaneous magnetization* $M_s(T)$ is attributed to exchange forces, which tend to aline neighboring electron spins parallel. Spin-orbit, quadrupole, and other interactions tend to orient the alined spin along particular crystalline axes; these interactions, in a phenomenological theory, are called forces of *(magneto)crystalline anisotropy*. Temperature agitation disturbs the alinement and causes $M_s(T)$ to decrease with increasing temperature, finally vanishing at the Curie point. An applied field $\boldsymbol{H}_0$ tends to aline $\boldsymbol{M}$ along it; it differs from the crystalline anisotropy forces in that opposite directions are equivalent with respect to them but not with respect to $\boldsymbol{H}_0$.

The effect of internal magnetic forces (dipole-dipole interactions) is the most difficult to predict. As was mentioned in the preceding section, the long-range part of the dipole-dipole energy is contained in either of the "magnetostatic self-energy" expressions W_m and W_m'. (The "local" part not taken into account in W_m or W_m' is one of the contributions to the crystalline anisotropy energy; the difference $W_m' - W_m$, by Eq. (2.54), reduces for a ferromagnetic body at uniform temperature to the constant $-\frac{1}{2}\gamma M_s^2 V$.) If the magnetization direction is constrained to be uniform, the tendency of magnetic forces is merely to rotate $\boldsymbol{M}$ into a direction of smallest W_m; for an ellipsoid, by Eq. (2.56), this is the direction of smallest demagnetizing factor, namely the longest principal axis. The concept of demagnetizing factor can be extended to any uniformly magnetized body (Brown and Morrish [*1*]; Brown [*10*], pp. 49—53), with a similar conclusion.

Experimentally, however, the magnetization is found not to be uniform except in very fine particles (of linear dimensions of order 10^2 Å or smaller). The reason can be found in Eq. (2.50), which expresses W_m in the form $(2\gamma)^{-1}\int \boldsymbol{H}_i^2\, dV$: this expression is nonnegative and would attain its minimum value if the magnetization could be arranged to be solenoidal, so that both $\nabla \cdot \boldsymbol{M}$ and $\boldsymbol{n} \cdot \boldsymbol{M}$ vanished; for then $\boldsymbol{H}_1$ would be zero. In a uniformly magnetized body, $\nabla \cdot \boldsymbol{M}$ indeed vanishes, but (except in an infinitely long cylinder) $\boldsymbol{n} \cdot \boldsymbol{M}$ cannot vanish everywhere on the surface, and the minimum W_{m} attainable by orientation is still positive. In a ring specimen, W_{m} can be made zero by directing $\boldsymbol{M}$ around the ring. For other shapes, W_{m} can usually not be reduced to zero except by discontinuous distributions, of which the more obvious ones give infinite exchange energy;[1] but suitably chosen nonuniform distributions can often produce a lower *total* energy than any uniform distribution.

The major competitors in this de-energization contest are the exchange forces, which favor uniformity, and the internal magnetic forces, which abhor poles. The former are much stronger than the latter in a struggle that involves nearest neighbors, but their strength falls rapidly to nothing as the distance increases, whereas the magnetic forces continue to be influential over surface-to-opposite-surface distances, even as the size of the specimen increases without limit (the factor D in Eq. (2.56) depends only on shape and not on size). The result is that the exchange forces win in fine particles but force a compromise in bodies of ordinary size: the magnetization in the latter case is nearly uniform over short distances, but its direction varies over larger ones.

The problem of finding the actual magnetization distribution is usually solved by the approximate methods of *domain theory* (Kittel [*1*]; Stewart [*1*]; Kittel and Galt [*1*]; Craik and Tebble [*1, 2*]; Träuble [*1*]), but some limited success has been achieved thru the rigorous approach of *micromagnetics* (Brown [*11*]; Shtrikman and Treves [*1*]; Kronmüller [*1*]).

For this problem, what is needed in addition to the theory of the previous sections is formal expressions for the exchange and anisotropy free-energy densities. To a sufficient approximation, the former can be expressed, at given temperature, as a constant plus a homogeneous nonnegative quadratic function of the spatial gradients of the direction cosines α_i of $\boldsymbol{M}$, invariant to a simultaneous rotation of $\boldsymbol{M}$ at all points (Landau and Lifshitz [*2*], p. 159); the latter is a function of these direction cosines themselves. Thus the "local" part of the free energy at constant T is, apart from a constant,

$$F_{\mathrm{loc}} = \int \{\tfrac{1}{2} b_{ij}\, \alpha_{k,i}\, \alpha_{k,j} + g(\alpha_1, \alpha_2, \alpha_3)\}\, d\tau, \tag{2.57}$$

[1] Certain line discontinuities (Brown [*11*], p. 97, footnote; Feldtkeller [*1*]) avoid both poles and infinite exchange energy, but their physical significance is questionable; at this point, physical realism requires that one abandon phenomenological theory and deal directly with such atomic concepts as an isolated line of reversed spins. The same is true of certain *point singularities* (Feldtkeller [*1*]).

with the quadratic form positive-definite.[1] For cubic crystals, if the cubic axes are chosen as coordinate axes,[2]

$$F_{\text{loc}} = \int \{\tfrac{1}{2} C \alpha_{i,j} \alpha_{i,j} + g(\alpha_1, \alpha_2, \alpha_3)\} d\tau \tag{2.58}$$

with $C > 0$ and with (KNELLER [*1*], Chap. 13)

$$g(\alpha_1, \alpha_2, \alpha_3) = K_1(\alpha_1^2 \alpha_2^2 + \alpha_2^2 \alpha_3^2 + \alpha_3^2 \alpha_1^2) + K_2 \alpha_1^2 \alpha_2^2 \alpha_3^2 + \cdots. \tag{2.59}$$

Because of the constraint $\alpha_i \alpha_i = 1$, such expressions are not unique; for example, the K_1 term is equivalent to $-\frac{1}{2} K_1 (\alpha_1^4 + \alpha_2^4 + \alpha_3^4)$ plus an additive constant (see also TUROV [*1*], Chap. 1).

In a small change $\delta\alpha_i$ (we suppose without loss of generality that $b_{ij} = b_{ji}$)

$$\left.\begin{aligned} \delta F_{\text{loc}} &= \int \left\{ b_{ij} \alpha_{k,i} \delta\alpha_{k,j} + \frac{\partial g}{\partial \alpha_k} \delta\alpha_k \right\} d\tau \\ &= \int \left\{ -b_{ij} \alpha_{k,ij} + \frac{\partial g}{\partial \alpha_k} \right\} \delta\alpha_k \, d\tau + \int b_{ij} \alpha_{k,i} \delta\alpha_k n_j \, dS. \end{aligned}\right\} \tag{2.60}$$

On the supposition that this is the only local energy and that the change is reversible, the above expression must be equal, by Eq. (2.53), to $\int \boldsymbol{H} \cdot \delta \boldsymbol{M} d\tau = \int M_s H_k \delta\alpha_k d\tau$. This holds for $\delta\alpha_k$'s subject to the constraint $\alpha_k \alpha_k = 1$, or $\alpha_k \delta\alpha_k = 0$, but otherwise arbitrary functions of position. We may therefore equate to zero the coefficient of $\delta\alpha_k$ in the difference $\delta F_{\text{loc}} - \int M_s H_k \delta\alpha_k d\tau - \int \lambda(x_i) \alpha_k \delta\alpha_k d\tau - \int \mu(x_i) \alpha_k \delta\alpha_k dS$, where λ and μ are Lagrangian multipliers. This gives in the volume V occupied by the specimen

$$-b_{ij} \alpha_{k,ij} + \frac{\partial g}{\partial \alpha_k} - M_s H_k - \lambda \alpha_k = 0 \tag{2.61}$$

[1] What is required physically is that $b_{ij} \alpha_{k,i} \alpha_{k,j}$ shall be positive for all values of the $\alpha_{k,i}$'s consistent with the constraint $\alpha_k \alpha_k = 1$. Positive definiteness of the quadratic form $b_{ij} x_i x_j$ is at least a *sufficient* condition for satisfaction of this requirement.

[2] The quantity here written $\frac{1}{2} C(\alpha_{i,j} \alpha_{i,j})$ was written $\frac{1}{2}\alpha[(\nabla s_x)^2 + (\nabla s_y)^2 + (\nabla s_z)^2]$ ("where s_x, s_y, s_z are the components of the magnetic moment $\boldsymbol{s}$ of unit volume") in 1935 by LANDAU and LIFSHITZ [*1*], Eq. (1); $(\alpha I^2/2)[(\nabla s_x)^2 + (\nabla s_y)^2 + (\nabla s_z)^2]$ ("where I is the magnetic saturation moment per unit volume, $\boldsymbol{s}$ is the unit vector in the direction of the magnetic moment") in 1944 by LIFSHITZ [*1*], Eq. (6); $\frac{1}{2}\alpha_1[(\partial \boldsymbol{M}/\partial x)^2 + (\partial \boldsymbol{M}/\partial y)^2] + \frac{1}{2}\alpha_2(\partial \boldsymbol{M}/\partial z)^2$, or more generally $\frac{1}{2}\alpha_{ik}(\partial M_l/\partial x_i)(\partial M_l/\partial x_k)$, in 1960 by LANDAU and LIFSHITZ [*2*], Eqs. (39.2) and (39.1). The symbol C for their αI^2 was first used in my 1940 paper on the approach to saturation (BROWN [*3*]). The symbol A for $\frac{1}{2}\alpha I^2 (= \frac{1}{2}C)$, used in a 1949 review by KITTEL [*1*], Eq. (2.1.11), has attained considerable currency; use of it is analogous to writing the kinetic energy of a particle kv^2 rather than $\frac{1}{2}mv^2$ (its equation of motion then becomes $F = 2ka$). Sometimes A is described as "the" exchange constant, with a reference to LANDAU and LIFSHITZ [*1*]; actually they never used that particular constant, and the reader who seeks a definition of it in their papers will seek in vain. In the present monograph, the generalization of $\frac{1}{2}C\alpha_{i,j}\alpha_{i,j}$ to general crystal symmetries will be written $\frac{1}{2}b_{ij}\alpha_{k,i}\alpha_{k,j}$, rather than $\frac{1}{2}C_{ij}\alpha_{k,i}\alpha_{k,j}$, because the symbol C_{AB} has been pre-empted in finite-strain theory [see § 3.2, Eq. (3.12)].

and on its bounding surface S

$$b_{ij}\alpha_{k,i}n_j - \mu\alpha_k = 0.^1 \qquad (2.62)$$

For cubic crystals ($b_{ij} = C\delta_{ij}$), the first terms in these two equations become, respectively, $-C\alpha_{k,ii} = -C\nabla^2\alpha_k$ and $C\alpha_{k,j}n_j = C\,\partial\alpha_k/\partial n$.

Equation (2.61) may be interpreted thus: for equilibrium, the "effective field"

$$\text{``}H_{\text{eff}\,k} = H_k + M_s^{-1}\{b_{ij}\alpha_{k,ij} + \partial g/\partial\alpha_k\}\text{''} \qquad (2.63)$$

must have zero component perpendicular to $\boldsymbol{M}$. The term "effective field" and the equation have been enclosed in quotation marks [as were Eqs. (2.42)—(2.43)] because their content is semantic rather than physical. One can return to physics by going over to *torque* equations: multiplication of Eqs. (2.61), (2.62) by α_m, followed by antisymmetrization with respect to k and m, eliminates λ and μ and gives equations that state that the torques per unit volume and per unit area that act on the unit vector $\boldsymbol{\alpha}$ must vanish. (These torques have no component along $\boldsymbol{\alpha}$.) The torque concept can be used directly if one wishes to extend the theory to dynamic processes.

Calculations based on these equations have been summarized elsewhere (BROWN [*11*], SHTRIKMAN and TREVES [*1*]). Part of our present problem is to modify them so as to include magnetoelastic interactions.

3. Concepts of Elasticity Theory

3.1. The analysis of stress. The classical analysis of stress has been presented fairly consistently, except for notation, in a number of treatises on elasticity; see, for example, LOVE [*1*] and SOKOLNIKOFF [*1*]. The object of the present discussion is to review the standard analysis, for an unpolarized and unmagnetized material, in such a way as to facilitate the transition to a magnetized material.

It is assumed that the body is subject to two types of force of *external* origin: *body forces* and *surface tractions*. The body force on volume element $d\tau$ is $\varrho\boldsymbol{f}\,d\tau$, where ϱ is the density and $\boldsymbol{f}$ is the force per unit mass; it is immaterial whether the force per unit volume is represented by the product $\varrho\boldsymbol{f}$ or by a single symbol. The surface traction on surface element dS of the body surface is $\boldsymbol{T}dS$. Both $\boldsymbol{T}$ and $\varrho\boldsymbol{f}$ may be supposed to be given functions, subject to direct experimental control. For example, $\varrho\boldsymbol{f}$ might be the gravitational force, which is at least closer to controllability in these days of lunar and Martian exploration than it was in the early days of elasticity theory; $\boldsymbol{T}$ might be a normal pressure $-\boldsymbol{n}p$ due to a surrounding fluid, or it might include tangential forces

[1] This equation requires additional terms if surface anisotropy is present (BROWN [*11*], Chaps. 3—4).

also. (We must then suppose that force-exerting machines act on the surface thru welded connections or thru layers of double-faced adhesive tape.)

In the standard theory, it is assumed (usually without explicit statement) that the only torques of external origin are the torques $\boldsymbol{r}\times(\varrho\boldsymbol{f})\,d\tau$ and $\boldsymbol{r}\times\boldsymbol{T}dS$; then the torque on the matter in an internal volume $\Delta\tau$, of given shape and of representative linear dimension L, about a point inside $\Delta\tau$ vanishes as L^4 when L approaches zero at constant shape. The transition to magnetic materials will be easier if we suppose that there may also be *body couples* of the form $\boldsymbol{C}\,d\tau$, where $\boldsymbol{C}$ is a vector function of position within the body; the contribution of $\boldsymbol{C}$ to the torque on $\Delta\tau$ vanishes only as L^3. The possible presence of such couples in a magnetized material is evident from Eqs. (2.11); it was mentioned by MAXWELL [*1*], Vol. II, pp. 276—279.

It is now postulated that the internal mutual forces between different parts of the body can be described as follows: across any internal surface element dS, with positive unit normal $\boldsymbol{n}$, there acts a force $\boldsymbol{t}(\boldsymbol{n})\,dS$ exerted by the matter on the positive side upon the matter on the negative side, together with an exactly opposite force exerted by the matter on the negative side upon the matter on the positive side. The *stress vector* $\boldsymbol{t}(\boldsymbol{n})$ is a function of position and of the direction of $\boldsymbol{n}$.

The standard steps in the analysis of stress are then the following.

First [see Fig. 11 (a)] we consider a small tetrahedron $\Delta\tau$ bounded by portions of three planes $x_i=\text{const}$ and by the plane triangle ΔS with vertices at $x_i+\lambda b_i$, with b_1, b_2, and b_3 positive constants and with λ a positive parameter that is to be decreased to zero. The equation of linear motion of the matter in the tetrahedron is, with omission of infinitesimals of higher order,

$$\varrho a_i\,\Delta\tau=\varrho f_i\,\Delta\tau+[t_i(\boldsymbol{n})-n_i t_{ij}]\,\Delta S, \tag{3.1}$$

where t_{ij} is the special value of $t_i(\boldsymbol{n})$ for a surface element with positive normal $\boldsymbol{n}$ along the x_j axis; $\boldsymbol{a}$ is the acceleration. As $\lambda\to 0$, $\Delta\tau$ vanishes as λ^3, ΔS as λ^2. If, therefore, we first divide by ΔS and then let $\lambda\to 0$, we get

$$t_i(\boldsymbol{n})=t_{ij}n_j. \tag{3.2}$$

The essence of this result is that it enables us to find $\boldsymbol{t}(\boldsymbol{n})$ for arbitrary $\boldsymbol{n}$, at a given point, if we know its values for three mutually orthogonal directions of $\boldsymbol{n}$. Its correctness depends on the assumption that an internal region ΔV is subject to no forces of forms other than those considered: namely, the body forces $\boldsymbol{f}$ and the stress vectors $\boldsymbol{t}(\boldsymbol{n})$.

Second [Fig. 11 (b)], we consider a small cube of edge L, with faces parallel to the coordinate axes. In this case we write the *angular* equations of motion and let $L \to 0$. The rate of change of angular momentum about the center of the cube is equal to the torque about this point. The x_1 component of the torque is $\int \varrho (x_2 f_3 - x_3 f_2)\, d\tau + L \cdot (t_{32} - t_{23}) \cdot L^2 + C_1 L^3$ plus higher-order infinitesimals. As $L \to 0$, the integral vanishes as

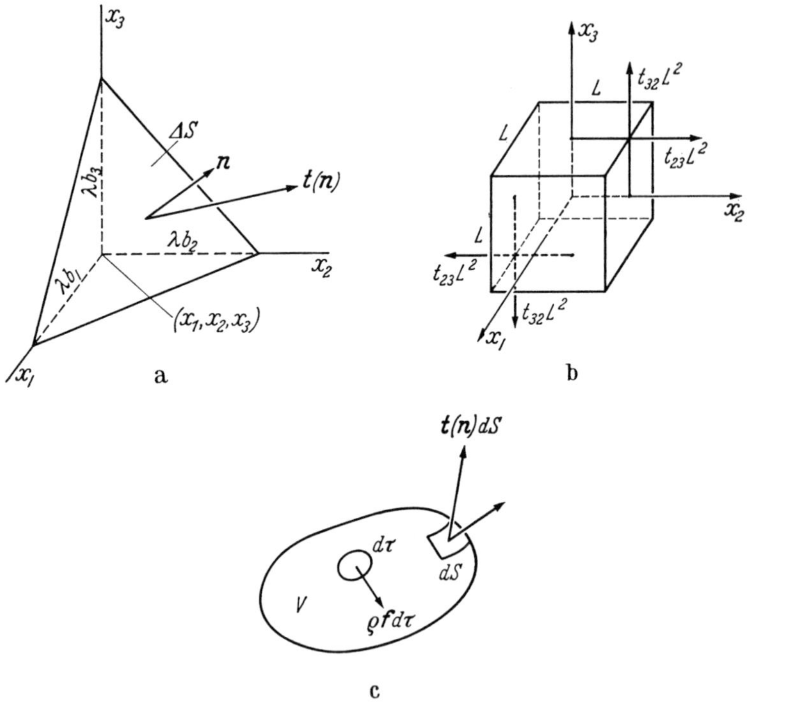

Fig. 11 (a)—(c). Geometries used in the standard analysis of stress. (a) A tetrahedron bounded by three planes $x_i = \text{const}$ and a plane area ΔS with normal $\boldsymbol{n}$; this leads to Eq. (3.2). (b) A cube bounded by planes $x_i = \text{const}$; this leads to Eq. (3.3). (c) An arbitrary region V; this leads to Eq. (3.5)

L^4, as does the angular-momentum term $(d/dt) \int \varrho (x_2 v_3 - x_3 v_2)\, d\tau$ ($\boldsymbol{v}$ = velocity). If, therefore, we first divide by L^3 and then let $L \to 0$, we get three equations such as

$$t_{23} - t_{32} = C_1. \tag{3.3}$$

In the usual theory, $\boldsymbol{C} = 0$, and this step demonstrates that the matrix (t_{ij}) is symmetric; when $\boldsymbol{C} \neq 0$, Eq. (3.3) enables us to evaluate its antisymmetric part if we know the couple density $\boldsymbol{C}$.

Third [Fig. 11 (c)], we consider the equation of linear motion of the matter instantaneously in an arbitrary finite volume V:

$$\int \varrho a_i\, d\tau = \int \varrho f_i\, d\tau + \int t_{ij} n_j\, dS = \int \{\varrho f_i + t_{ij,j}\}\, d\tau. \tag{3.4}$$

Since this holds for an arbitrary V, we may equate the integrands. This gives the local equations of motion

$$\varrho a_i = \varrho f_i + t_{ij,j}. \tag{3.5}$$

Fourth, we consider the equations of angular motion of the matter in V. This step, with use of the previous relations, leads to an identity. If we wish, we may use this step instead of step 2 to derive Eq. (3.3).

Finally, to find boundary conditions at an external surface of the body, we may consider the motion of a very thin layer of matter just inside a surface element ΔS. On the assumption that the internal forces do not vary in a special and sudden manner as the surface is approached, we find that the surface traction $\boldsymbol{T}\Delta S$ of external origin must balance the internal force $-\boldsymbol{t}(\boldsymbol{n})\Delta S$ due to matter farther in; thus at the surface

$$T_i = t_{ij} n_j. \tag{3.6}$$

In a rotation of coordinates $x^*_\alpha = l_{\alpha i} x_i$ and $x_i = l_{\alpha i} x^*_\alpha$ ($l_{\alpha i} l_{\alpha j} = \delta_{ij}$, $l_{\alpha i} l_{\beta i} = \delta_{\alpha\beta}$), we have for the transformed vector $\boldsymbol{t}^*(\boldsymbol{n}^*)$

$$t^*_i(\boldsymbol{n}^*) = t_{ij} n^*_j = t_{ij} n^*_\beta l_{\beta j}; \tag{3.7}$$

hence

$$t^*_\alpha(\boldsymbol{n}^*) = l_{\alpha i} t^*_i(\boldsymbol{n}^*) = l_{\alpha i} t_{ij} n^*_\beta l_{\beta j} \equiv t^*_{\alpha\beta} n^*_\beta, \tag{3.8}$$

with

$$t^*_{\alpha\beta} = l_{\alpha i} l_{\beta j} t_{ij}. \tag{3.9}$$

By setting n^*_β in Eq. (3.8) equal to $\delta_{\beta\gamma}$, we see that $t^*_{\alpha\gamma}$ is the special value of $t^*_i(\boldsymbol{n}^*)$ for a surface element with positive normal $\boldsymbol{n}^*$ along the x_γ axis; that is, $t^*_{\alpha\beta}$ has the same physical meaning in the x^*_α axes that t_{ij} has in the x_i axes. Eq. (3.9) therefore shows that t_{ij}, defined on the basis of this physical meaning, transforms as a tensor of rank 2.

3.2. Finite strains. We need the following concepts of finite-strain theory (Truesdell and Toupin [*1*]).

Suppose that the particle originally at $\boldsymbol{R}$ or X_A ($A = 1, 2, 3$) is, after deformation of the body by the forces $\varrho \boldsymbol{f} d\tau$ and $\boldsymbol{T} dS$, at $\boldsymbol{r}$ or x_i ($i = 1, 2, 3$). Then the functions $x_i(X_A)$ or the inverse functions $X_A(x_i)$ describe the deformation for the body as a whole, and the nine partial derivatives $x_{i,A} \equiv \partial x_i/\partial X_A$ or $X_{A,i} \equiv \partial X_A/\partial x_i$ describe it locally. These derivatives satisfy the relations

$$x_{i,A} X_{A,j} = \delta_{ij}, \qquad X_{A,i} x_{i,B} = \delta_{AB}, \tag{3.10}$$

since the left members are merely circuitous paraphrases of $\partial x_i/\partial x_j$ and $\partial X_A/\partial X_B$. The functions $x_i(X_A)$ are more convenient in finite-strain theory than are the *displacements*

$$u_i \equiv x_i - \delta_{iA} X_A; \tag{3.11}$$

when $u_i = 0$ (or a constant with respect to the variables X_A), $x_i = \delta_{iA} X_A$.[1,2]

The finite *strain tensor* E_{AB} is defined by

$$C_{AB} \equiv \delta_{AB} + 2E_{AB} \equiv x_{i,A}\, x_{i,B}. \tag{3.12}$$

Its claim to be considered a measure of strain is based on the expressions for the square of the distance between two neighboring points before and after the deformation,

$$d s_0^2 = dX_A\, dX_A \tag{3.13}$$

and

$$d s^2 = dx_i\, dx_i = x_{i,A}\, x_{i,B}\, dX_A\, dX_B = C_{AB}\, dX_A\, dX_B; \tag{3.14}$$

when $E_{AB} = 0$, $C_{AB} = \delta_{AB}$ and $d s^2 = d s_0^2$, *i.e.*, the neighborhood under consideration has undergone at most a rigid-body displacement and rotation. Eq. (3.14) shows that the quadratic form on the right is positive definite.

The tensor C_{AB} is symmetric; it therefore represents only 6 independent quantities instead of the original 9 quantities $x_{i,A}$. Being symmetric, C_{AB} can be diagonalized by proper choice of axes. Let the diagonal elements in this representation [the principal values of the tensor, which by Eq. (3.14) are positive] be C_1, C_2, C_3. Then the symmetric tensor with the same principal axes but with elements $C_1^{\frac{1}{2}}$, $C_2^{\frac{1}{2}}$, $C_3^{\frac{1}{2}}$ may be defined as the tensor $(C^{\frac{1}{2}})_{AB}$; it is at once seen that $(C^{\frac{1}{2}})^2 = C$, so that, in any axes, $(C^{\frac{1}{2}})_{AC}(C^{\frac{1}{2}})_{CB} = C_{AB}$. The local deformation[3]

$$\bar{x}_{i,A} = (C^{\frac{1}{2}})_{iA} \tag{3.15}$$

gives strain tensor $\bar{E}_{AB}$, where

$$\left.\begin{aligned} \bar{C}_{AB} &= \delta_{AB} + 2\bar{E}_{AB} = \bar{x}_{i,A}\bar{x}_{i,B} = (C^{\frac{1}{2}})_{iA}(C^{\frac{1}{2}})_{iB} \\ &= (C^{\frac{1}{2}})_{Ai}(C^{\frac{1}{2}})_{iB} = C_{AB}. \end{aligned}\right\} \tag{3.16}$$

---

[1] The distinction between capital and lower-case subscripts can no longer be preserved as $x_{i,A} \to \delta_{iA}$, *i.e.*, in the small-displacement approximation; there, however, it ceases to be important.

[2] We regard the functions $x_i(X_A)$ as describing a *deformation* even in the limiting case $x_1 = X_1$, $x_2 = X_2$, $x_3 = X_3$ (or $x_i = \delta_{iA} X_A$), when the *displacements* u_i are zero. Consistently, we shall call the quantities $x_{i,A}$ *deformation gradients*; they differ from the *displacement gradients* $u_{i,A}$ by δ_{iA}. Thus when the displacement vector vanishes, the *displacement gradient* tensor also vanishes, but the *deformation gradient* tensor reduces to the identity tensor δ_{iA}.

[3] The formal distinction between capital and lower-case subscripts has here been abandoned in $(C^{\frac{1}{2}})_{iA}$. It could be preserved by writing the right member $\delta_{iC}(C^{\frac{1}{2}})_{CA}$. The distinction is unimportant, however, except after a comma: $\bar{x}_{i,A}$ means $\partial\bar{x}_i/\partial X_A$, whereas $\bar{x}_{i,j}$ would mean $\partial\bar{x}_i/\partial x_j$; $\bar{x}_{1,2}$ is ambiguous and must be replaced by the less compact $\partial\bar{x}_1/\partial X_2$.

Thus the deformation $\bar{x}_{i,A}$ results in the same strains (that is, in the same changes of interparticle distances ds) as the deformation $x_{i,A}$ but is described by a symmetric tensor. We may now define a tensor R_{Ai} by

$$R_{Ai} \equiv x_{i,B}(C^{-\frac{1}{2}})_{AB}, \tag{3.17}$$

where $(C^{-\frac{1}{2}})_{AB}$ is defined by the same method already used to define $(C^{\frac{1}{2}})_{AB}$; it is easily seen that, as the notation implies, $C^{-\frac{1}{2}}$ is the inverse of $C^{\frac{1}{2}}$. Operation on Eq. (3.17) with $(C^{\frac{1}{2}})_{CA}$ therefore gives

$$(C^{\frac{1}{2}})_{CA} R_{Ai} = x_{i,B}\delta_{CB} = x_{i,C} \tag{3.18}$$

or

$$x_{i,A} = (C^{\frac{1}{2}})_{AB} R_{Bi}. \tag{3.19}$$

From Eq. (3.17) we find directly

$$\left.\begin{aligned} R_{Ai} R_{Ci} &= x_{i,B} x_{i,D} (C^{-\frac{1}{2}})_{AB} (C^{-\frac{1}{2}})_{CD} = (C^{-\frac{1}{2}})_{AB} C_{BD} (C^{-\frac{1}{2}})_{DC} \\ &= (C^{-\frac{1}{2}})_{AB} (C^{\frac{1}{2}})_{BE} (C^{\frac{1}{2}})_{ED} (C^{-\frac{1}{2}})_{DC} = \delta_{AE}\delta_{EC} = \delta_{AC}, \end{aligned}\right\} \tag{3.20}$$

so that R_{Ai} is an orthogonal tensor. Its determinant must be either $+1$ or -1; when $x_{i,B} = \delta_{iB}$, $C_{AB} = \delta_{AB}$, therefore $(C^{-\frac{1}{2}})_{AB} = \delta_{AB}$ and, by Eq. (3.17), $R_{Ai} = \delta_{iA}$, with determinant 1; by continuity, $\det R_{Ai} = 1$ always. Thus R_{Ai} describes a rigid-body rotation of the neighborhood under consideration, and Eq. (3.19) analyzes the deformation into a finite rotation followed by a finite strain. The three Euler angles equivalent to R, plus the six components of C, are together equivalent to the nine components of $x_{i,A}$.

A mass element dm occupies before the distortion a volume $d\tau_0 = dX_1 dX_2 dX_3$, and after it a new volume $d\tau = dx_1 dx_2 dx_3$. The relation between the two is

$$d\tau = J\, d\tau_0, \tag{3.21}$$

with

$$J = \frac{\partial(x_1, x_2, x_3)}{\partial(X_1, X_2, X_3)} = \det x_{i,A}. \tag{3.22}$$

The densities before and after are ϱ_0 and ϱ respectively, with $\varrho\, d\tau = \varrho_0\, d\tau_0$, so that

$$\varrho = J^{-1} \varrho_0. \tag{3.23}$$

These formulas enable us to change volume integrals over the deformed coordinates x_i into volume integrals over the undeformed coordinates X_A. In order to deal also with surface integrals, we need the relation of the deformed surface element dS, with normal $\boldsymbol{n}$, to the undeformed element dS_0, with normal $\boldsymbol{N}$ (TRUESDELL and TOUPIN [1], p. 203):

$$dS = J\left[(C^{-1})_{AB} N_A N_B\right]^{\frac{1}{2}} dS_0. \tag{3.24}$$

Eq. (3.24) can be derived as follows. Let the undeformed surface be $\boldsymbol{R} = \boldsymbol{R}(\lambda, \mu)$ or $X_A = X_A(\lambda, \mu)$, where λ and μ are parameters. Then an element of undeformed vector area is $d\boldsymbol{S}_0 = d\boldsymbol{R}^{(1)} \times d\boldsymbol{R}^{(2)}$, or

$$dS_{0A} = e_{ABC}\, dX_B^{(1)}\, dX_C^{(2)}, \tag{3.25}$$

where

$$dX_B^{(1)} = (\partial X_B/\partial \lambda)\, d\lambda, \qquad dX_C^{(2)} = (\partial X_C/\partial \mu)\, d\mu,$$

and e_{ABC} is 1 if A, B, C are an even permutation of 1, 2, 3, -1 if they are an odd permutation, and 0 otherwise.[1] The corresponding element of deformed area is $d\boldsymbol{S} = d\boldsymbol{r}^{(1)} \times d\boldsymbol{r}^{(2)}$, or

$$\left.\begin{aligned} dS_p &= e_{pqr}\, dx_q^{(1)}\, dx_r^{(2)} = e_{pqr}\, x_{q,A}\, x_{r,B}\, dX_A^{(1)}\, dX_B^{(2)} \\ &= \frac{\partial(x_q, x_r)}{\partial(X_A, X_B)}\, dX_A^{(1)}\, dX_B^{(2)}; \end{aligned}\right\} \tag{3.26}$$

in the last expression, A and B are still summed over, but p, q, r are an even permutation of $1, 2, 3$: $q = p+1 \pmod 3$ and $r = p+2 \pmod 3$. The next steps enable us to express the right member of Eq. (3.26) in terms of the undeformed elements dS_{0A}. By the first Eq. (3.10), $x_{p,P}\, x_{P,q} = \delta_{pq}$. If, for given q, these three equations ($p = 1, 2, 3$) are solved for the three quantities $X_{1,q}$, $X_{2,q}$, and $X_{3,q}$, the result is

$$X_{P,p} = J^{-1}\, \partial(x_q, x_r)/\partial(X_Q, X_R), \tag{3.27}$$

where both p, q, r and P, Q, R are even permutations of 1, 2, 3. In Eq. (3.26), where p, q, r are such a permutation, $\partial(x_q, x_r)/\partial(X_A, X_B)$ is therefore equal to $J X_{C,p}$ when A, B, C are also such a permutation; it is easily seen to be $-J X_{C,p}$ when they are an odd permutation of 1, 2, 3 and to be zero when $B = A$. Thus Eq. (3.26) may be written

$$dS_p = e_{ABC}\, J X_{C,p}\, dX_A^{(1)}\, dX_B^{(2)} = J X_{C,p}\, dS_{0C}, \tag{3.28}$$

by Eq. (3.25). This is the desired modification of Eq. (3.26). We now have

$$dS^2 = dS_p\, dS_p = J^2 X_{C,p}\, X_{D,p}\, dS_{0C}\, dS_{0D}. \tag{3.29}$$

Now

$$X_{C,p}\, X_{D,p} = (C^{-1})_{CD}, \tag{3.30}$$

as can be shown by postmultiplying the left member by $C_{DE} = x_{i,D}\, x_{i,E}$ and using Eq. (3.12). Thus, finally,

$$dS^2 = J^2 (C^{-1})_{CD}\, dS_{0C}\, dS_{0D} = J^2 (C^{-1})_{CD}\, N_C\, N_D\, dS_0^2, \tag{3.31}$$

and on taking the square root we get Eq. (3.24).

---

[1] When general curvilinear coordinates are used, it is necessary to distinguish between two sets of quantities, e_{ijk} and ε_{ijk}, that differ by a factor independent of the subscripts. Unfortunately there is no consistency between different authors about which set shall be represented by e and which by ε. In Cartesian coordinates the two fortunately are identical.

From Eq. (3.28),

$$n_p = dS_p/dS = JX_{C,p}\, dS_{0C}/dS. \tag{3.32}$$

Eq. (3.24) can be derived more easily, but without derivation of Eq. (3.32), by the following method. Choose temporary $\bar{X}_A$ axes to coincide with the principal axes of the tensor C after $\boldsymbol{dS}$ has undergone the rotation R (which does not affect dS). In these axes, the remaining deformation consists of extensions in ratios $C_1^{\frac{1}{2}}$, $C_2^{\frac{1}{2}}$, and $C_3^{\frac{1}{2}}$ respectively along the $\bar{X}_1$, $\bar{X}_2$, and $\bar{X}_3$ axes. The original projected area $dS_{01} = \bar{N}_1\, dS_0$ perpendicular to the $\bar{X}_1$ axis therefore becomes $d\bar{S}_1 = C_2^{\frac{1}{2}} C_3^{\frac{1}{2}} \bar{N}_1\, dS_0$, and so on.

Hence

$$\left.\begin{aligned} dS^2 &= d\bar{S}_1^2 + d\bar{S}_2^2 + d\bar{S}_3^2 = (C_2 C_3 \bar{N}_1^2 + C_3 C_1 \bar{N}_2^2 + C_1 C_2 \bar{N}_3^2)\, dS_0^2 \\ &= J^2 (\bar{N}_1^2/C_1 + \bar{N}_2^2/C_2 + \bar{N}_3^2/C_3)\, dS_0^2, \end{aligned}\right\} \tag{3.33}$$

since $C_1^{\frac{1}{2}} C_2^{\frac{1}{2}} C_3^{\frac{1}{2}} = d\tau/d\tau_0 = J$, so that $C_1 C_2 C_3 = J^2$. On transforming to arbitrary X_A axes, we get Eq. (3.31).

3.3. Energy and stress-strain relations. In this section we suppose, as is usual in elasticity theory, that there are no body couples: $\boldsymbol{C} = 0$ in Eq. (3.3).[1] We now calculate the rate of work by all the forces exerted on the matter instantaneously occupying a volume V of the specimen, by sources outside V:

$$\frac{\delta W}{\delta t} = \int \varrho f_i v_i\, d\tau + \int_{S_1} T_i v_i\, dS + \int_{S_2} t_i v_i\, dS. \tag{3.34}$$

Here $\boldsymbol{v}$ is the velocity of the particle instantaneously at x_i; when we write it thus, we shall suppose that it is expressed as a function of time t and of the deformed coordinates x_i rather than of the undeformed coordinates X_A: $v_i = v_i(x_j, t)$. The surface bounding V may include a part S_1 that is part of the bounding surface of the specimen and a part S_2 that lies inside the specimen.

We now eliminate $\boldsymbol{f}$, $\boldsymbol{T}$, and $\boldsymbol{t}$ by use of Eqs. (3.2), (3.5), and (3.6), then transform the surface integral into a volume integral:

$$\left.\begin{aligned} \frac{\delta W}{\delta t} &= \int (\varrho a_i - t_{ij,j})\, v_i\, d\tau + \int t_{ij} v_i n_j\, dS \\ &= \int \dot{x}_i \ddot{x}_i\, dm - \int t_{ij,j}\, v_i\, d\tau + \int (t_{ij} v_i)_{,j}\, d\tau \\ &= \frac{d}{dt}\left\{\frac{1}{2}\int \dot{x}_i^2\, dm\right\} + \int t_{ij} v_{i,j}\, d\tau; \end{aligned}\right\} \tag{3.35}$$

[1] This enables us to evade the problem of calculating the rate of work by body couples. In the magnetic case that will later be our primary object of interest, the couples are of magnetic origin, and the method of calculation is such that explicit consideration of the couples is not necessary (§ 6.1).

in the first term, we have expressed the velocity $\boldsymbol{v}$ and acceleration $\boldsymbol{a}$ as functions of time and of the undeformed coordinates X_A: $v_i(x_i, t) = (\partial/\partial t)\, x_i(X_A, t) \equiv \dot{x}_i(X_A, t)$, $a_i = \ddot{x}_i(X_A, t)$. This term is clearly the rate of change of kinetic energy $\mathcal{T}$. We therefore have

$$\frac{\delta W}{\delta t} - \frac{d\mathcal{T}}{dt} = \int t_{ij} v_{i,j}\, d\tau. \tag{3.36}$$

But

$$v_{i,j} = \dot{x}_{i,A} X_{A,j}, \tag{3.37}$$

and therefore

$$\frac{\delta W}{\delta t} - \frac{d\mathcal{F}}{dt} = \int t_{ij} \dot{x}_{i,A} X_{A,j}\, d\tau. \tag{3.38}$$

The right member of Eq. (3.38) evaluates the excess of the work rate of the external forces over the part accounted for as rate of increase of kinetic energy. We now suppose that this is accounted for as local stored energy (by "local" we imply mutual energy between particles very close together on the usual phenomenological or *megascopic* scale of the continuum theory), which can therefore be written as the mass integral of an energy per unit mass F. We also suppose that at given temperature, F depends only on the nine deformation gradients $x_{i,A}$. Then

$$\frac{d}{dt}\int \mathsf{F}\, dm = \int \frac{\partial \mathsf{F}}{\partial x_{i,A}} \dot{x}_{i,A}\, dm = \int \varrho \frac{\partial \mathsf{F}}{\partial x_{i,A}} \dot{x}_{i,A}\, d\tau. \tag{3.39}$$

If F is to account for all the nonkinetic energy,

$$\int t_{ij} \dot{x}_{i,A} X_{A,j}\, d\tau = \int \varrho \frac{\partial \mathsf{F}}{\partial x_{i,A}} \dot{x}_{i,A}\, d\tau. \tag{3.40}$$

The equality (3.40) holds for an arbitrary volume within the specimen, and therefore the integrands may be equated. This gives, at any point considered, $(t_{ij} X_{A,j} - \varrho\, \partial\mathsf{F}/\partial x_{i,A})\, \dot{x}_{i,A} = 0$; and since we can give the nine quantities $\dot{x}_{i,A}$ *at this one point* arbitrary values, their coefficients must vanish:

$$t_{ij} X_{A,j} = \varrho \frac{\partial \mathsf{F}}{\partial x_{i,A}}. \tag{3.41}$$

Operation on each member with $x_{k,A}$ gives

$$t_{ik} = \varrho \frac{\partial \mathsf{F}}{\partial x_{i,A}} x_{k,A}. \tag{3.42}$$

By symmetry, the energy of a mass element is not affected by a rigid rotation of the element.[1] Therefore if we go over from the 9 variables

[1] To a physicist, it would seem more natural to impose this condition *before* carrying out the differentiation that led to Eq. (3.39). The reversing of the natural order, however, simplifies the mathematics, especially in the later treatment of magnetizable materials. Cf. Truesdell ([*1*], pp. 174—175); Toupin ([*1*], p. 884).

$x_{i,A}$ to the 6 variables $E_{AB}=E_{BA}$ and the 3 variables equivalent to the rotation R, F will be independent of these last. If

$$\mathsf{F}=\mathscr{F}(E_{PQ}) \qquad (P, Q=1, 2, 3) \tag{3.43}$$

and if we treat E_{PQ} and E_{QP} as independent variables in the formal differentiations, then

$$\left.\begin{aligned}
\frac{\partial \mathsf{F}}{\partial x_{i,A}} &= \frac{\partial \mathscr{F}}{\partial E_{PQ}}\frac{\partial E_{PQ}}{\partial x_{i,A}} = \frac{1}{2}\frac{\partial \mathscr{F}}{\partial E_{PQ}}\frac{\partial C_{PQ}}{\partial x_{i,A}}\\
&= \frac{1}{2}\frac{\partial \mathscr{F}}{\partial E_{PQ}}\frac{\partial}{\partial x_{i,A}}(x_{j,P}x_{j,Q}) = \frac{1}{2}\frac{\partial \mathscr{F}}{\partial E_{PQ}}(\delta_{ij}\,\delta_{AP}\,x_{j,Q}+x_{j,P}\,\delta_{ij}\,\delta_{AQ})\\
&= \frac{1}{2}\frac{\partial \mathscr{F}}{\partial E_{PQ}}(\delta_{AP}\,x_{i,Q}+\delta_{AQ}\,x_{i,P}) = \frac{1}{2}\left(\frac{\partial \mathscr{F}}{\partial E_{AQ}}x_{i,Q}+\frac{\partial \mathscr{F}}{\partial E_{PA}}x_{i,P}\right)\\
&= \frac{1}{2}\left(\frac{\partial \mathscr{F}}{\partial E_{AP}}+\frac{\partial \mathscr{F}}{\partial E_{PA}}\right)x_{i,P}.
\end{aligned}\right\} \tag{3.44}$$

Hence

$$t_{ik}=\frac{1}{2}\varrho\left(\frac{\partial \mathscr{F}}{\partial E_{AP}}+\frac{\partial \mathscr{F}}{\partial E_{PA}}\right)x_{i,P}\,x_{k,A}. \tag{3.45}$$

This relates the stresses t_{ik} to the strains E_{AP}. Interchange of the free indices i and k will lead to the same expression, if the dummy indices are interchanged; thus $t_{ki}=t_{ik}$, in accordance with the assumed absence of body couples.

The derivation of Eqs. (3.42) and (3.45) depends on the postulate that the internal forces $\boldsymbol{t}\,dS$ are actually localized at the surface element dS. Only then is the rate of work by them correctly given by the third term in Eq. (3.34). A weaker postulate (TRUESDELL [*1*]) is that "upon any imagined closed surface with unit normal $\boldsymbol{n}$ within a body there exists a distribution of *stress* vectors $\boldsymbol{t}(\boldsymbol{n})$ whose resultant and moment are equivalent respectively to those of the actual forces of material continuity exerted by the material outside upon the material inside". If we use only this postulate, we can still derive all the formulas of § 3.3, since only these particular properties of $\boldsymbol{t}(\boldsymbol{n})$ were used there. But we cannot justify the third term of Eq. (3.34) and therefore cannot derive Eq. (3.40) except when V is the whole volume of the specimen. We therefore cannot pass from Eq. (3.40) to Eq. (3.42).

If we adopt this weaker postulate, then Eq. (3.40) holds for the specimen volume V. If we replace $\dot{x}_{i,A}\,X_{A,j}$ in the left member by the original v_i and replace $\dot{x}_{i,A}$ in the right member by its equivalent $v_{i,j}\,x_{j,A}$, we get

$$\int P_{ij}\,v_{i,j}\,d\tau=0, \tag{3.46}$$

where

$$P_{ij}\equiv t_{ij}-\varrho\frac{\partial \mathsf{F}}{\partial x_{i,A}}x_{j,A}; \tag{3.47}$$

the integration extends over the whole specimen. The equation remains true if we multiply by δt and replace $v_i \delta t$ by a virtual variation $\delta x_i(x_j)$. The functions $\delta x_{i,j}$, however, are not independent, since there are nine of them, whereas the functions that are actually arbitrary are the three functions δx_i; the $\delta x_{i,j}$'s are subject to compatibility conditions (TRUESDELL [1], p. 145). We therefore may not set P_{ij} equal to zero. Instead, we may transform as follows:[1]

$$0 = \int P_{ij}\, \delta x_{i,j}\, d\tau = \int P_{ij}\, \delta x_i\, n_j\, dS - \int P_{ij,j}\, \delta x_i\, d\tau. \tag{3.48}$$

The δx_i's are arbitrary functions of x_1, x_2, x_3; therefore we may equate their coefficients to zero. This gives $P_{ij,j}=0$ in V and $P_{ij} n_j = 0$ on S. Thus if t_{ij} differs from $\varrho(\partial \mathsf{F}/\partial x_{i,A})\, x_{j,A}$, it differs from it by a tensor P_{ij} that does not occur either in the internal equations of motion (3.5) or in the boundary conditions (3.6), when these are expressed in terms of the $x_{i,A}$'s by replacing t_{ij} by $\varrho(\partial \mathsf{F}/\partial x_{i,A})\, x_{j,A} + P_{ij}$. Therefore the values of the x_i's found by solving Eqs. (3.5) and (3.6) are independent of P_{ij}. We conclude that if the weaker postulate about t_i is adopted, there is an indeterminacy in the "stresses" t_{ij}, but the indeterminacy is without physical significance;[2] as far as any observable quantities are concerned, we may resolve the indeterminacy arbitrarily,[3] and most simply by setting $P_{ij}=0$.[4]

We shall encounter similar indeterminacies when we consider magnetic elastic solids.

3.4. Infinitesimal strains. The commonest form of linear elasticity theory (LOVE [1], SOKOLNIKOFF [1]) assumes that the displacements $u_i = x_i - \delta_{iA} X_A$ are small and vary slowly with position, so that all the deformation gradients $x_{i,A}$ are small. It is possible to make less restric-

[1] We are here regarding the virtual variations δx_i as functions of the *deformed* coordinates x_j rather than of the *undeformed* coordinates X_A. The latter method can also be used; but the mathematics of the present method is simpler.

[2] One can measure a force by balancing it against a known force, or by observing the acceleration it produces. The latter method yields only a value of the right member of Eq. (3.5) and hence, if $\boldsymbol{f}$ is known, a value of $t_{ij,j}$. The former method, within the limitations of the present (nonmagnetic) model, yields only the values of the controlled forces $\boldsymbol{f}$ and $\boldsymbol{T}$ necessary to satisfy Eq. (3.5) inside the body and Eq. (3.6) on its surface; therefore it yields only the values of $t_{ij,j}$ in the body and of $t_{ij} n_j$ on its surface, and not the values of t_{ij} itself.

[3] It is easily verified that the stress vector $t_i'(\boldsymbol{n}) = P_{ij} n_j$ gives zero force and moment on an arbitrary closed surface within a region where $P_{ij,j} = 0$. If, therefore, $t_i(\boldsymbol{n}) = t_{ij} n_j$ gives the correct forces and moments, so also does $t_i(\boldsymbol{n}) + t_i'(\boldsymbol{n})$. The condition $P_{ij} n_j = 0$ on the specimen surface S insures that addition of the term P_{ij} does not disturb the boundary conditions.

[4] For the parallelopiped $0 \leqq x_i \leqq a_i$, a particular nonzero solution of the equations $P_{ij,j} = 0$ in V, $P_{ij} n_j = 0$ on S, and $P_{ij} = P_{ji}$ is $P_{11} = \partial^2\chi/\partial x_2^2$, $P_{12} = P_{21} = -\partial^2\chi/\partial x_1 \partial x_2$, $P_{22} = \partial^2\chi/\partial x_1^2$, other P_{ij}'s $= 0$, with $\chi = -A \sin(n_1 \pi x_1/a_1) \times \sin(n_2 \pi x_2/a_2)$; $A = $ const, n_1 and n_2 = positive integers.

tive assumptions, for example to allow a finitely bent rod with small strains, but we shall consider only the usual theory.

The approximations made consist in omitting terms of higher than the second order of small quantities in the energy, and terms of higher than the first in the formulas for stresses as functions of the $x_{i,A}$'s. There are two aspects of this truncation process: truncation of the energy as a function of the strains, and truncation of the strains and rotations as functions of the displacement gradients

$$u_{i,A} = x_{i,A} - \delta_{iA}. \tag{3.49}$$

We consider the second of these first.

From the definitions (3.12) and (3.11),

$$\left.\begin{aligned} E_{AB} &= \tfrac{1}{2}(C_{AB} - \delta_{AB}) = \tfrac{1}{2}(x_{i,A}\, x_{i,B} - \delta_{AB}) \\ &= \tfrac{1}{2}[(\delta_{iA} + u_{i,A})(\delta_{iB} + u_{i,B}) - \delta_{AB}] \\ &= \tfrac{1}{2}(u_{A,B} + u_{B,A}) + \tfrac{1}{2} u_{i,A}\, u_{i,B} = u_{(A,B)} + \tfrac{1}{2} u_{i,A}\, u_{i,B}; \end{aligned}\right\} \tag{3.50}$$

this is exact. To the second order in $u_{i,A}$,

$$C^{-\frac{1}{2}} = (1 + 2E)^{-\frac{1}{2}} = 1 - E + \tfrac{3}{2} E^2, \tag{3.51}$$

or

$$(C^{-\frac{1}{2}})_{AB} = \delta_{AB} - u_{(A,B)} - \tfrac{1}{2} u_{i,A}\, u_{i,B} + \tfrac{3}{2} u_{(A,C)}\, u_{(C,B)}, \tag{3.52}$$

whence by Eq. (3.17)

$$\left.\begin{aligned} R_{Ai} = x_{i,B}(C^{-\frac{1}{2}})_{AB} &= (\delta_{iB} + u_{i,B})(C^{-\frac{1}{2}})_{AB} = \delta_{Ai} + \tfrac{1}{2}(u_{i,A} - u_{A,i}) - \\ &- \tfrac{1}{2}[u_{j,A}\, u_{j,i} + u_{i,B}(u_{A,B} + u_{B,A})] + \\ &+ \tfrac{3}{8}(u_{A,C} + u_{C,A})(u_{C,i} + u_{i,C}). \end{aligned}\right\} \tag{3.53}$$

The distinction between small and capital letters has now become blurred. However,

$$\partial u_i/\partial X_A = (\partial u_i/\partial x_j)\, x_{j,A} = (\partial u_i/\partial x_j)(\delta_{jA} + u_{j,A}) = \partial u_i/\partial x_A$$

to the first order; therefore if $u_{i,j}$ occurs either in a first-order (stress) formula, or multiplied by another first-order quantity in a second-order (energy) formula, it may be interpreted at pleasure as $\partial u_i/\partial X_j$ or as $\partial u_i/\partial x_j$. Under these conditions, the first-order formulas

$$E_{AB} = u_{(A,B)}, \qquad R_{Ai} = \delta_{Ai} + u_{[i,A]} \tag{3.54}$$

are sufficient. The analysis (3.19) of the deformation $x_{i,A}$ into a finite rotation R_{Bi} followed by a finite strain C_{AB} reduces, in the first-order approximation, to the usual separation of the tensor $u_{i,A}$ into its symmetric part $u_{(i,A)}$ (the infinitesimal strain tensor) and its antisymmetric part $u_{[i,A]}$ (the infinitesimal rotation tensor):

$$u_{i,A} = u_{(i,A)} + u_{[i,A]} = \tfrac{1}{2}(u_{i,A} + u_{A,i}) + \tfrac{1}{2}(u_{i,A} - u_{A,i}) = E_{iA} + \Omega_{iA}. \tag{3.55}$$

The small rigid rotation $\boldsymbol{\Omega} \cdot d\boldsymbol{r}$ can also be written $\boldsymbol{\omega} \times \boldsymbol{r}$, where $\boldsymbol{\omega}$ is the axial vector with components $\omega_1 = -\Omega_{23} = +\Omega_{32}$ etc.

We now suppose that F can be expanded as a TAYLOR's series in the E_{AB}'s about the value at $E_{AB} = 0$, and that terms of order 3 and higher may be neglected. Then

$$\mathsf{F} = \mathsf{F}^{(0)} + \mathsf{F}_{AB} E_{AB} + \tfrac{1}{2} \mathsf{F}_{ABCD} E_{AB} E_{CD}. \tag{3.56}$$

We may without loss of generality suppose that $\mathsf{F}_{AB} = \mathsf{F}_{BA}$ and that $\mathsf{F}_{ABCD} = \mathsf{F}_{BACD} = \mathsf{F}_{ABDC}$. Then Eq. (3.45) gives

$$\left.\begin{aligned} t_{ik} &= \frac{1}{2} \varrho \left(\frac{\partial \mathsf{F}}{\partial E_{AB}} + \frac{\partial \mathsf{F}}{\partial E_{BA}} \right) x_{i,B} x_{k,A} \\ &= \varrho (\mathsf{F}_{AB} + \mathsf{F}_{ABCD} E_{CD}) (\delta_{iB} + u_{i,B}) (\delta_{kA} + u_{k,A}), \end{aligned}\right\} \tag{3.57}$$

which in the "undeformed" state ($u_{i,P} = 0$) reduces to $\varrho_0 \mathsf{F}_{ki} (= \varrho_0 \mathsf{F}_{ik})$. We define the "undeformed state" at a given temperature T as the equilibrium state under zero external body and surface forces ($\boldsymbol{f} = 0, \boldsymbol{T} = 0$). The equilibrium equations (3.5) (with $a_i = 0$) and (3.6) then require that

$$(\varrho_0 \mathsf{F}_{ik})_{,k} = 0 \quad \text{in } V, \qquad \varrho_0 \mathsf{F}_{ik} n_k = 0 \quad \text{on } S. \tag{3.58}$$

In a homogeneous material, the first of these conditions is satisfied by any constants F_{ik}. The second condition must also be satisfied for a specimen of arbitrary shape, with values of the F_{ik}'s that depend only on the material and not on the shape. For a cube with edges along the coordinate axes, the conditions on the three pairs of surfaces lead to the requirement $\mathsf{F}_{i1} = \mathsf{F}_{i2} = \mathsf{F}_{i3} = 0$ for any $i (= 1, 2, 3)$. Hence $\mathsf{F}_{AB} = 0$, and to the second order, by Eq. (3.56),

$$\mathsf{F} = \mathsf{F}^{(0)} + \tfrac{1}{2} \mathsf{F}_{ABCD} E_{AB} E_{CD}; \tag{3.59}$$

to the first order, by Eq. (3.57),

$$t_{ik} = \varrho \mathsf{F}_{kiCD} E_{CD} = \varrho_0 \mathsf{F}_{iklm} u_{(l,m)}. \tag{3.60}$$

We may replace E_{AB} by $u_{(A,B)}$ in Eq. (3.59), since it is multiplied by the first-order small quantity E_{CD}, and we may replace E_{CD} by $u_{(C,D)}$. At the same time we may define a new energy function F by

$$F = \varrho_0 \mathsf{F} = F^{(0)} + \tfrac{1}{2} F_{ijkl} u_{i,j} u_{k,l}, \tag{3.61}$$

where $F^{(0)} = \varrho_0 \mathsf{F}^{(0)}$ and $F_{ijkl} = \varrho_0 \mathsf{F}_{ijkl}$; F is the energy per unit undistorted volume. Then

$$t_{ik} = F_{iklm} u_{(l,m)} = \frac{1}{2} \left[\frac{\partial F}{\partial u_{l,m}} + \frac{\partial F}{\partial u_{m,l}} \right]. \tag{3.62}$$

Insertion of this value of t_{ik} into the equations of motion (3.5) and boundary conditions (3.6), together with replacement of ϱ by ϱ_0, gives the usual equations of linear elasticity. The last substitution is justified because if Eq. (3.5) is written $a_i = f_i + \varrho^{-1} t_{ij,j}$, then since t_{ij} is a first-order small quantity, replacement of ϱ^{-1} by ϱ_0^{-1} introduces first-order errors in the factors ϱ^{-1} but only second-order errors in the term $\varrho^{-1} t_{ij,j}$.

The foregoing discussion justifies the usual approximations of linear elasticity theory: in particular, the use of $u_{(i,j)}$ as a measure of strain and the differentiation of an energy per unit volume to get stress formulas [Eq. (3.62)]. It should be noticed, however, that the replacement of E_{AB} by $u_{(A,B)}$ and of ϱ by ϱ_0 would *not* be justified in a term in F of the form $\mathsf{F}_{AB} E_{AB}$. In magnetostriction theory, such terms are present; in fact, in the conventional theory of magnetostriction they are the *only* sources of magnetoelastic interaction.

Most presentations of linear elasticity theory introduce $u_{(i,j)}$ as a measure of strain at the very beginning. It is then very difficult to be sure of the correctness of any extension of the theory, *e.g.* to magnetoelastic interactions.

4. Thermodynamic Principles

4.1. Introductory remarks. If there is an area of physics that has slipped over into metaphysics, it is thermodynamics. Not quantum mechanics: the mysticism there is precisely measured out in small parcels of amount $\hbar$; the mysticism in thermodynamics shrouds the foundations of the subject in an impenetrable fog. For some remarks on specific examples of this, see TRUESDELL [*1*], p. 161. Another example is the following. In mechanics, we define work done by and against *forces*, and we have equality of action and reaction in the sense that $\boldsymbol{F}_{21} = -\boldsymbol{F}_{12}$ [as in Eq. (2.1)]. In thermodynamics, we discover that these concepts have been stealthily replaced by the notions of work δW_{21} done *on a system by its surroundings*, and work δW_{12} done *on the surroundings by the system*, and we are told that $\delta W_{21} = -\delta W_{12}$. Suppose that the "system" is a charge q_1, the "surroundings" a charge q_2. Is δW_{21} the work done by $\boldsymbol{F}_{21}$ when q_2 causes q_1 to move? Then δW_{12} must be the work done by $\boldsymbol{F}_{12}$ when q_1 causes q_2 to move, and $\delta W_{21} \neq -\delta W_{12}$ unless the charges move in unison. Or is δW_{21} the work done against $\boldsymbol{F}_{12}$ when q_2 succeeds in moving? Then δW_{12} must be the work done against $\boldsymbol{F}_{21}$ when q_1 succeeds in moving; and again $\delta W_{21} \neq -\delta W_{12}$. It appears that traditional thermodynamics is tacitly excluding all interactions except those that occur by surface contact, so that $\delta \boldsymbol{r}_1 = \delta \boldsymbol{r}_2$. The exclusion may be a legitimate step to limit the area of discussion; the taciturnity about it is not.

Obscurity of this sort seems quite unnecessary; the concept of work as defined in mechanics should still be applicable to bodies undergoing temperature changes, and it applies as well to an elastic solid as to a fluid. On the other hand, the obscurity connected with the concept of entropy seems to be inherent in the concept itself. In order to talk about the entropy change that occurs in the smashing of a mirror, we must be able to define the entropy before and after the smashing; and according to the traditional definition, this means that we must find a reversible process by which to get from the unsmashed to the smashed state — in other words, we must be able to unsmash the mirror. To avoid the difficulties of this situation, some authors resort to a postulational method of developing the subject; it then becomes a mathematically perfect set of abstract propositions with no obvious relation to anything physical.

Resolution of these difficulties will not be attempted here. Magnetic hysteresis is not as drastic a process as mirror-smashing; in the simple models that have been examined in detail, the required reversible paths in fact exist. The traditional definition of entropy therefore suffices for our purposes. The object of the present discussion is to state the traditional theory in a form directly applicable to the magnetizable elastic bodies with which we shall be concerned. In that discussion, however, there will be no work done on or by systems; work will be done, as in mechanics, by or against forces.

4.2. The first and second laws. If a system is described by (generalized) coordinates q_α, and if the (generalized) force of external origin that tends to increase q_α is P_α, the rate of work by these forces is $P_\alpha \dot{q}_\alpha$; the work by them in a small displacement δq_α is $P_\alpha \delta q_\alpha$. If, simultaneously, an amount of heat δQ enters the body from the surroundings, the first law of thermodynamics states that there is a function U of the coordinates (and also, perhaps, of other parameters Θ_β) such that, in *any* change,

$$\delta U = P_\alpha \delta q_\alpha + \delta Q; \tag{4.1}$$

U is the *internal energy*. The second law states that there is another function η, the entropy, such that, in *any* change, by suitable definition of the (Kelvin) temperature T,

$$\delta Q \leqq T \delta \eta; \tag{4.2}$$

the equality holds in the ideal limiting case of a *reversible* process; the inequality holds in all other cases. On eliminating δQ, we get

$$\delta U \leqq P_\alpha \delta q_\alpha + T \delta \eta. \tag{4.3}$$

4.3. Equilibrium and stability conditions. We consider first reversible processes; and more specifically, processes in which the system is in equilibrium (or at least arbitrarily close to equilibrium) at each stage. In such a process,

$$\delta U = P_\alpha \delta q_\alpha + T \delta \eta. \tag{4.4}$$

It is clear that U is not a function of the q_α's alone, but also of η (and perhaps of other parameters Θ_β that remain constant during the changes being considered). If we know U as a function of the q_α's and η, we can find the P_α's and the temperature, as functions of the same variables, by using the relations

$$P_\alpha = \partial U/\partial q_\alpha, \qquad T = \partial U/\partial \eta. \tag{4.5}$$

We seldom have such knowledge; in particular, we are likely to know the temperature rather than the entropy. For this case, let

$$F = U - T\eta; \tag{4.6}$$

then in any reversible change,

$$\delta F = \delta U - T\,\delta\eta - \eta\,\delta T = P_\alpha\,\delta q_\alpha - \eta\,\delta T. \tag{4.7}$$

If we know F as a function of the q_α's and T, we can find the P_α's and η by using the relations

$$P_\alpha = \partial F/\partial q_\alpha, \qquad \eta = -\,\partial F/\partial T. \tag{4.8}$$

Often we wish to use the P_α's, rather than the q_α's, as independent variables. For this case, let

$$G = F - P_\alpha\, q_\alpha: \tag{4.9}$$

then in any reversible change

$$\delta G = \delta F - P_\alpha\,\delta q_\alpha - q_\alpha\,\delta P_\alpha = -\,q_\alpha\,\delta P_\alpha - \eta\,\delta T. \tag{4.10}$$

If we know G as a function of the P_α's and T, we can find the q_α's and η by using the relations

$$q_\alpha = -\,\partial G/\partial P_\alpha, \qquad \eta = -\,\partial G/\partial T. \tag{4.11}$$

If we wish to use a combination of q_α's and P_α's as independent variables, we extend the sum in Eq. (4.9) only over those α's for which the independent variable is P_α; thus for independent variables T, q_1, and P_2, the appropriate function is $F - P_2\, q_2$, and its differential is $P_1\,\delta q_1 - q_2\,\delta P_2 - \eta\, dT$. Similarly, if the independent variables are η, q_1, and P_2, we use $U - P_2\, q_2$, whose differential is $P_1\,\delta q_1 - q_2\,\delta P_2 + T\,\delta\eta$. By such *Legendre transformations* we can get the proper *thermodynamic potential* for any choice of independent variables; all the dependent variables can be found by differentiation of this one function.

In traditional thermodynamics courses, it is usually assumed that the system under consideration is fluid and that the only $P_\alpha\,\delta q_\alpha$ term needed is $-p\,\delta V$ (p = pressure, V = volume). A considerable amount of the course is devoted to working out all the possible relations among all possible partial derivatives for all possible choices of the independent variables: for example, V and η, V and T, p and η, and p and T. The special names *Helmholtz function*, *enthalpy*, and *Gibbs function* are attached to the thermodynamic potentials $F = U - T\eta$, $H = U + pV$, and $G = U - T\eta + pV$. Mnemonic devices are presented to help the student remember such relations as $\partial p/\partial T = \partial\eta/\partial V$ — relations that he would have no need to remember if he knew only the first and second laws, the procedures illustrated above for changing the independent variables, and the condition for a perfect differential. Most of this traditional thermodynamics is useless for our purposes. We shall not attempt to give names to all the thermodynamic potentials we encounter; we shall, however, apply the term *free energy* to those that use T as independent variable, and the term *energy* to those that use η.

The theory based on Eq. (4.4) enables us to derive relations between the dependent and independent variables for a system in equilibrium, *i.e.* to derive equilibrium conditions; but it tells us nothing about the stability of the equilibrium. To treat this problem, we return to the form of (4.3) that holds in naturally occurring processes,

$$\delta U < P_\alpha\,\delta q_\alpha + T\,\delta\eta. \tag{4.12}$$

As it stands, this inequality tells us that if we hold the coordinates and the entropy constant ($\delta q_\alpha = 0$ and $\delta\eta = 0$), the internal energy U can only decrease; and that if we hold the coordinates and the energy constant, the entropy can only increase. The condition for stable equilibrium is therefore that the energy already be as small as it can be for the given coordinate values and entropy, or that the entropy already be as large as it can be for the given coordinate values and energy. Only the latter situation is experimentally realizable (by holding the coordinates constant and preventing heat flow between the body and its surroundings); under these conditions, we find stable equilibrium states by *maximizing* the entropy with respect to any parameters on which it depends (these will also determine the P_α's and T).

If, instead, we hold the q_α's and T constant, (4.12) can be rewritten (since $\delta q_\alpha = 0$)

$$\delta(U - T\eta) < 0 \tag{4.13}$$

or

$$\delta F < 0. \tag{4.14}$$

By the same reasoning as before, we find that for stable equilibrium at given q_α's and T, the thermodynamic potential F must be a minimum with respect to any parameters on which it depends.

If we hold the P_α's and T constant, (4.12) can be rewritten

$$\delta(U - T\eta - P_\alpha\,q_\alpha) < 0 \tag{4.15}$$

or

$$\delta G < 0. \tag{4.16}$$

For stable equilibrium at given P_α's and T, G must be a minimum. Similarly, if we hold q_1, P_2, and T constant, the condition for stable equilibrium is that $F - P_2 q_2$ must be a minimum.

Thus for any set of controlled conditions, we can state as the condition for stable equilibrium that a certain thermocynamic potential must be a minimum. The proper thermodynamic potential is the one whose independent variables in the reversible relations are the variables held constant in the search for equilibrium.

The condition for mere *equilibrium*, without regard to stability, is the vanishing of the first variation of the appropriate thermodynamic potential. Thus for equilibrium at given q_α's and T, $\delta F = 0$; for equilibrium at given P_α's and T, $\delta G = 0$, or

$$\delta F - P_\alpha\, \delta q_\alpha = 0. \tag{4.17}$$

Formally, this is Eq. (4.7) with $\delta T = 0$. In Eq. (4.7), however, the variations were actual variations, with the system in equilibrium at each stage; in Eq. (4.17), they are virtual variations from the equilibrium state, and they include arbitrary variations of internal parameters that were fully determined in Eq. (4.7).

4.4. Application to rigid magnetic bodies. For a rigid magnetizable body, we have seen (§ 2.6) that the work done by the forces due to the magnetizing coil, when the magnetization changes, is $\int \boldsymbol{H}_0 \cdot \delta \boldsymbol{M}\, \delta\tau$; thus $P_\alpha = H_{0i}(\boldsymbol{r})$, $q_\alpha = M_i(\boldsymbol{r})\, d\tau$, and the sum over α becomes a sum over the Cartesian components ($i = 1, 2, 3$) and an integral over the volume occupied by the specimen. Therefore Eq. (4.7), which holds in reversible changes, becomes

$$\delta F = \int \boldsymbol{H}_0 \cdot \delta \boldsymbol{M}\, d\tau - \eta\, \delta T; \tag{4.18}$$

in a reversible isothermal process, $\delta F = \int \boldsymbol{H}_0 \cdot \delta \boldsymbol{M}\, d\tau$. If we set $\boldsymbol{H}_0 = \boldsymbol{H} - \boldsymbol{H}_1$, we can (cf § 2.6) transform this to $\delta(F - W_{\mathrm{m}}) = \int \boldsymbol{H} \cdot \delta \boldsymbol{M}\, d\tau$, where W_{m} is one possible form of the "magnetic self-energy" and includes the long-range, shape-dependent part of the dipole-dipole energy; then $F_{\mathrm{loc}} \equiv F - W_{\mathrm{m}}$ is a part of F that can be represented as the volume integral of a free-energy density, dependent only on local conditions. In this section we shall use as independent variable $\boldsymbol{H}_0$, which is subject to direct control, rather than $\boldsymbol{H}$, which is not; and we shall take account of W_{m} by including it explicitly in F. The thermodynamic potential appropriate to $\boldsymbol{H}_0$ and T as independent variables is

$$G \equiv F - \int \boldsymbol{H}_0 \cdot \boldsymbol{M}\, d\tau; \tag{4.19}$$

its differential in a reversible change is

$$\delta G = -\int \boldsymbol{M} \cdot \delta \boldsymbol{H}_0\, d\tau - \eta\, \delta T; \tag{4.20}$$

the condition for stable equilibrium at given $\boldsymbol{H}_0$ and T is that G be a minimum; and the condition for equilibrium, without regard to stability, at given $\boldsymbol{H}_0$ and T is that $\delta G=0$ for arbitrary virtual variations of the parameters on which G depends.

On using the defining equation (4.19) and on inserting the known long-range part of F, we have

$$G=F_{\mathrm{loc}}+W_{\mathrm{m}}-\int \boldsymbol{H}_0\cdot\boldsymbol{M}\,d\tau=F_{\mathrm{loc}}-\tfrac{1}{2}\int \boldsymbol{H}_1\cdot\boldsymbol{M}\,d\tau-\int \boldsymbol{H}_0\cdot\boldsymbol{M}\,d\tau. \qquad (4.21)$$

By use of Eq. (2.57) for F_{loc}, this becomes, for a ferromagnetic body,

$$G=\int\{\tfrac{1}{2}b_{ij}\,\alpha_{k,i}\,\alpha_{k,j}+g(\alpha_m)\}\,d\tau-\tfrac{1}{2}\int \boldsymbol{H}_1\cdot\boldsymbol{M}\,d\tau-\int \boldsymbol{H}_0\cdot\boldsymbol{M}\,d\tau, \qquad (4.22)$$

where $M_i=M_{\mathrm{s}}\,\alpha_i$, with M_{s} constant. The first variation of G is (again we take $b_{ij}=b_{ji}$)

$$\left.\begin{aligned}\delta G=\int\{b_{ij}\,\alpha_{k,i}\,\delta\alpha_{k,j}+(\partial g/\partial\alpha_k)\,\delta\alpha_k\}\,d\tau-\\ -\tfrac{1}{2}\int(\boldsymbol{H}_1\cdot\delta\boldsymbol{M}+\boldsymbol{M}\cdot\delta\boldsymbol{H}_1)\,d\tau-\int \boldsymbol{H}_0\cdot\delta\boldsymbol{M}\,d\tau.\end{aligned}\right\} \qquad (4.23)$$

We integrate the first term by parts to get rid of the derivatives of variations, $\delta\alpha_{k,j}$; and in the magnetostatic term, we use the reciprocity relation (2.49) to replace $\int \boldsymbol{M}\cdot\delta\boldsymbol{H}_1\,d\tau$ by $\int \boldsymbol{H}_1\cdot\delta\boldsymbol{M}\,d\tau$. This gives, since $\boldsymbol{H}_1+\boldsymbol{H}_0=\boldsymbol{H}$,

$$\left.\begin{aligned}\delta G=\int b_{ij}\,\alpha_{k,i}\,\delta\alpha_k\,n_j\,dS+\\ +\int\{-b_{ij}\,\alpha_{k,ij}+(\partial g/\partial\alpha_k)-M_{\mathrm{s}}H_k\}\,\delta\alpha_k\,d\tau.\end{aligned}\right\} \qquad (4.24)$$

To take account of the constraint $\alpha_k\,\alpha_k=1$, or $\alpha_k\,\delta\alpha_k=0$, we require that $\delta G-\int\lambda\alpha_k\,\delta\alpha_k\,d\tau-\int\mu\alpha_k\,\delta\alpha_k\,dS=0$, where λ and μ are Lagrangian multipliers (functions of position), with the $\delta\alpha_k$'s now arbitrary functions of position. On equating the coefficient of $\delta\alpha_k$ to zero, we again get Eqs. (2.61) and (2.62).

The test for stability of the equilibrium requires investigation of the second variation of G. This investigation has been carried out only in a few special cases (Brown [*11*], pp. 72—91).

We now examine a simpler material; the results will cast some doubt on our thermodynamic method. We suppose that for a volume element with a specified magnetization, and in internal equilibrium but not necessary in equilibrium with the fields that act upon it,

$$F_{\mathrm{loc}}=\int\frac{1}{2\chi}\,\boldsymbol{M}^2\,d\tau, \qquad (4.25)$$

where χ is independent of $\boldsymbol{M}$ but may be a function of position. Then

$$G=\int\frac{1}{2\chi}\,\boldsymbol{M}^2\,d\tau-\frac{1}{2}\int \boldsymbol{H}_1\cdot\boldsymbol{M}\,d\tau-\int \boldsymbol{H}_0\cdot\boldsymbol{M}\,d\tau, \qquad (4.26)$$

$$\delta G=\int\left\{\frac{1}{\chi}\,\boldsymbol{M}-\boldsymbol{H}_1-\boldsymbol{H}_0\right\}\cdot\delta\boldsymbol{M}\,d\tau; \qquad (4.27)$$

and by use of the alternative expression $\frac{1}{2\gamma}\int \boldsymbol{H}_1^2\,d\tau$ for W_{m},

$$\delta^2 G = \int \frac{1}{2\chi}(\delta \boldsymbol{M})^2\,d\tau + \frac{1}{2\gamma}\int (\delta \boldsymbol{H}_1)^2\,d\tau. \tag{4.28}$$

The equilibrium condition $\delta G = 0$, for arbitrary $\delta \boldsymbol{M}$, gives $\boldsymbol{M} = \chi(\boldsymbol{H}_1 + \boldsymbol{H}_0) = \chi \boldsymbol{H}$; thus the material is magnetically linear and isotropic, with susceptibility χ. The stability condition $\delta^2 G > 0$ (the condition that the equilibrium G be a minimum) is satisfied, according to Eq. (4.28), if χ is everywhere positive. What, then, about diamagnetic materials?

The substitution $\boldsymbol{H}_1 = \boldsymbol{B}_1 - \gamma \boldsymbol{M}$ changes Eq. (4.26) to

$$G = \int \frac{1}{2}\left(\frac{1}{\chi} + \gamma\right) \boldsymbol{M}^2\,d\tau - \frac{1}{2}\int \boldsymbol{B}_1 \cdot \boldsymbol{M}\,d\tau - \int \boldsymbol{H}_0 \cdot \boldsymbol{M}\,d\tau; \tag{4.29}$$

the second term is W'_{m} of § 2.6. On setting $\delta G = 0$, we get a relation equivalent to $\boldsymbol{M} = \chi \boldsymbol{H}$; but the new formula for the second variation is

$$\delta^2 G = \int \frac{1}{2}\left(\frac{1}{\chi} + \gamma\right)(\delta \boldsymbol{M})^2\,d\tau - \frac{1}{2\gamma}\int (\delta \boldsymbol{B}_1)^2\,d\tau. \tag{4.30}$$

This is *negative* if $\frac{1}{\chi} + \gamma$ is negative, *i.e.* if χ is negative and smaller in absolute value than $1/\gamma$. The range $0 > \chi > -1/\gamma$ is precisely the range in which the susceptibilities of diamagnetic materials lie. According to our thermodynamic theory, therefore, diamagnetic materials are magnetically unstable and cannot exist.

If one examines this problem microscopically, one soon finds what was wrong. The magnetic moment of a particle, when expressed in terms of coordinates and *momenta* rather than of coordinates and *velocities*, contains an intrinsic diamagnetic (negative) term that is not subject to arbitrary variation; it is, by definition, determined by the magnetic field and incapable of being out of equilibrium with the field. If we subtract out this term in the magnetization and use only the rest in our thermodynamic theory, we get results consistent with experiments. But thermodynamics is supposedly a very general phenomenological theory whose validity does not rest on microscopic models of the phenomena!

For our purposes, fortunately, this is an academic matter: in ferromagnetic materials, we may safely neglect the very small intrinsic diamagnetic part of the magnetization.

4.5. Application to elastic nonmagnetic bodies. For the whole of a nonmagnetic elastic body, by Eq. (3.34), the work done by the body forces $\boldsymbol{f}$ and surface tractions $\boldsymbol{T}$ in a small change is $\int \varrho f_i\,\delta x_i\,d\tau + \int T_i\,\delta x_i\,dS$. Here δx_i is an actual or virtual variation of the position of the particle whose position before the variation is x_i; δx_i may be expressed as a function either of the undeformed coordinates X_A or of the deformed coordinates x_j. The sum over α, in the general formulas of § 4.3, becomes a sum over Cartesian components i and an integral over the volume and surface. Thus Eq. (4.7) for reversible changes becomes

$$\delta F = \int \varrho f_i \, \delta x_i \, d\tau + \int T_i \, \delta x_i \, dS - \eta \, \delta T; \tag{4.31}$$

in a reversible isothermal process, $\delta F = \int \varrho f_i \, \delta x_i \, d\tau + \int T_i \, \delta x_i \, dS$. For natural changes, the basic inequality (4.12) becomes

$$\delta U < \int \varrho f_i \, \delta x_i \, d\tau + \int T_i \, \delta x_i \, dS + T \delta\eta; \tag{4.32}$$

and for changes at constant T, since then $\delta U - T\delta\eta = \delta(U - T\eta) = \delta F$,

$$\delta F - \int \varrho f_i \, \delta x_i \, d\tau - \int T_i \, \delta x_i \, dS < 0. \tag{4.33}$$

To do a Legendre transformation to the external forces as independent variables, we must decide what quantities related to these forces are to be held constant during the variations δx_i. One possibility is to keep the force on each particle constant, regardless of its position. With the undeformed coordinates as variables of integration, Eq. (4.33) becomes

$$\delta F - \int \varrho_0 f_i \, \delta x_i \, d\tau_0 - \int T_i' \, \delta x_i \, dS_0 < 0, \tag{4.34}$$

where $T_i' \equiv (dS/dS_0)\, T_i$ is the force per unit undistorted area. For natural changes in which f_i and T_i' remain constant functions of the undeformed coordinates X_A,

$$\delta\{F - \int \varrho_0 f_i x_i \, d\tau_0 - \int T_i' x_i \, dS_0\} < 0; \tag{4.35}$$

the thermodynamic potential to be minimized under these conditions, in order to find a stable equilibrium state, is therefore

$$G_1 \equiv F - \int \varrho_1 f_i x_i \, d\tau_0 - \int T_i' x_i \, dS_0. \tag{4.36}$$

A situation easier to realize experimentally is that $\boldsymbol{f} = -\nabla\psi$, i.e. $f_i = -\psi_{,i} = -\partial\psi/\partial x_i$ (*not* $\partial\psi/\partial X_i$!), where $\psi(x_i)$ is potential energy per unit mass (*e.g.* a gravitational potential), and that $\boldsymbol{T} = -\boldsymbol{n}p$, where p is a uniform and constant hydrostatic pressure exerted by a surrounding fluid. In this case, the inequality (4.33) becomes

$$\delta F + \int \delta\boldsymbol{r} \cdot \nabla\psi \, dm + p \int \boldsymbol{n} \cdot \delta\boldsymbol{r} \, dS < 0 \tag{4.37}$$

or (V is the specimen volume)

$$\delta F + \delta \int \psi dm + p\,\delta V < 0. \tag{4.38}$$

Here the ψ associated with each particle is the value at the instantaneous position of that particle; ψ remains constant at any point fixed in space and varies only because the particles move. At given T and p, with ψ constant at every fixed point in space,

$$\delta\{F+\int\psi\, dm+pV\}<0; \tag{4.39}$$

the thermodynamic potential to be minimized is therefore

$$G=F+\int\psi\, dm+pV. \tag{4.40}$$

The problem is simpler if $\boldsymbol{f}$ and $\boldsymbol{T}$ vanish. Then, according to inequality (4.33), we must minimize F with respect to the functions $x_i(X_A)$; the boundary conditions are those that come out of the minimization process itself. If $\boldsymbol{f}$ vanishes and the surface values of x_i are prescribed, then by (4.33) we again minimize F, but this time under the constraints x_i = prescribed value on S; these constraints now provide the boundary conditions.

The problem is also simpler if we are interested only in finding the equilibrium state and not in testing its stability. Then we may use Eq. (4.17), which becomes (4.33) with $<$ replaced by $=$, and with δx_i now an arbitrary virtual variation. In this case, if we set $F=\int \mathsf{F}(x_{i,A})\,dm$ as in § 3.3, we have

$$\left.\begin{aligned}\delta F&=\int\frac{\partial \mathsf{F}}{\partial x_{i,A}}\,\delta x_{i,A}\,\varrho_0\,d\tau_0=\int\frac{\partial \mathsf{F}}{\partial x_{i,A}}\,\delta x_{i,j}\,x_{j,A}\,\varrho\,d\tau\\&=\int\varrho\frac{\partial \mathsf{F}}{\partial x_{i,A}}\,x_{j,A}\,\delta x_i\,n_j\,dS-\int\left(\varrho\frac{\partial \mathsf{F}}{\partial x_{i,A}}\,x_{j,A}\right)_{,j}\delta x_i\,d\tau.\end{aligned}\right\} \tag{4.41}$$

We may now equate to zero the coefficients of the arbitrary δx_i's, in V and on S, in the expression for $\delta F-\int\varrho f_i\,\delta x_i\,d\tau-\int T_i\,\delta x_i\,dS$. This gives

$$\left(\varrho\frac{\partial \mathsf{F}}{\partial x_{i,A}}\,x_{j,A}\right)_{,j}-\varrho f_i=0 \quad\text{in}\quad V \tag{4.42}$$

and

$$\varrho\frac{\partial \mathsf{F}}{\partial x_{i,A}}\,x_{j,A}\,n_j-T_i=0 \quad\text{on}\quad S. \tag{4.43}$$

If we set

$$t_{ij}=\varrho\frac{\partial \mathsf{F}}{\partial x_{i,A}}\,x_{j,A}, \tag{4.44}$$

Eqs. (4.42) and (4.43) become identical with the previous equilibrium equations (3.5) (with $a_i=0$) and (3.6). Furthermore, Eq. (4.44) is identical with the previous formula (3.42) for t_{ij}. But Eqs. (4.42), (4.43) would still be satisfied if we defined t_{ij} as the expression (4.44) plus P_{ij}, where P_{ij} satisfies the same conditions as in § 3.3: namely, $P_{ij,j}=0$ in V and $P_{ij}\,n_j=0$ on S. Thus the thermodynamic minimization method leads to the same indeterminacy in the t_{ij} formula as does the stress method (augmented by an elementary energy-conservation argument) of §§ 3.1 and 3.3, if we adopt the weaker of the two postulates about the stress vector $\boldsymbol{t}(\boldsymbol{n})$ [see the discussion following Eq. (3.45)].

Chapter II

Force and Stress Relations in a Deformable Magnetic Material

5. The Forces in General

5.1. Prolog. In the preceding sections, we developed separately the theory of a rigid magnetized body and the theory of a nonmagnetic elastic solid. In each case two methods were used. In the first method (§§ 2 and 3), field and stress concepts played a dominant role; energy considerations entered only toward the end (§§ 2.6 and 3.3) and only thru an elementary "conservation of energy" argument. In the second method (§§ 4.4 and 4.5), energy considerations were the basis of the treatment; the key principle was minimization of a thermodynamic potential. The concept of internal mutual forces, or stresses, entered only at the end and as a triviality of notation [Eq. (4.44)]; the concept of an internal self-field $\boldsymbol{H}_1$ entered only in the magnetostatic "self-energy" W_m, where it could in fact have been avoided by substituting for $\boldsymbol{H}_1$ the expression (with appropriate subscript changes) (2.35). The results by the two methods were equivalent, provided the conserved energy of the first treatment was, for isothermal changes, identified with the "free energy" (Helmholtz function) F of the second.

We shall now carry out analogous treatments of a body that is both magnetizable and deformable. In this chapter, the field-and-stress method will be used; in Chapter III, the thermodynamic minimization method. The latter has, in principle at least, the advantage that it provides not only equilibrium equations but a stability test (positiveness of the second variation of the thermodynamic potential G).

In § 2.7, the special properties of a ferromagnetic material, viewed on a microscopic but not atomic scale, were discussed; for a rigid material, these properties could be summarized as (1) constancy of the magnitude of the magnetization and (2) presence of a free-energy term quadratic in the spatial gradients of the direction cosines of the magnetization. Extension of these concepts to a deformable material will be left till Chapter III. They are more easily handled by the energy method; furthermore, in the present chapter our main object is to investigate carefully the effects of magnetization on elastic concepts and theorems and of strain on magnetic concepts and theorems, and for this purpose it is preferable to avoid introducing simultaneously the complications due to exchange forces. The theory of this chapter will therefore apply to a paramagnetic material, without restriction to the magnetically linear range, or to a ferromagnetic material viewed on a megascopic scale (so that domain structure is not observable). When energy relations are used, the ferromagnetic case must be restricted to a field range in which the behavior is nearly reversible; for single crystals, there is a considerable range in which this is true (BOZORTH [*1*], Chap. 12).

By disregarding the special properties of ferromagnets, we make the theory of magnetizable and deformable materials exactly analogous, mathematically, to the theory of deformable dielectrics. To obtain the latter from the former, we need only to make the substitutions $\boldsymbol{M} \to \boldsymbol{P}$ (electric moment per unit volume or polarization), $\boldsymbol{H} \to \boldsymbol{E}$ (electric "field intensity"), and $\boldsymbol{B} \to \boldsymbol{D}$ (electric "induction"). In our unit system, even the constant γ remains unchanged. Amperian currents in magnetostatics have their analog, tho not a frequently used one (moving magnetic poles; see BROWN [*2*]), in electrostatics. The possible electric interpretation of the equations will be useful in interpreting certain formulas, *viz.* those relating to poles, since the electric analog — electric charges — is physically less artificial.

The treatment of magnetostatics in §§ 2.1—2.4 requires no modification for the new problem; but all the coordinates in this treatment must be the instantaneous coordinates x_i of the particles *after* the deformation. The force and torque calculation of § 2.5 is also still valid as far as it goes; but to use it in the analysis of stress, we shall have to extend it so as to obtain formulas for the "magnetic" force and torque on a *part* of a body. In this extension we shall avoid postulates of the kind implied, tho not explicitly stated, in the treatments by MAXWELL and his editors, discussed in connection with "Eq." (2.42). The work calculation in § 2.6 must be redone for the case of a deformable body.

The analysis of stress in § 3.1 requires important revisions for a magnetizable (or electrically polarizable) body; the finite-strain analysis of § 3.2 is independent of the possible presence of polarization, but the work and energy calculations of § 3.3 require revision.

We have found (at the end of § 3.3) that even in an unpolarizable material, the "stresses" cannot be defined uniquely, if in our definition of the stress vector $\boldsymbol{t}$ we avoid superfluous concepts by specifying only properties that in principle are observable. We shall now find that this principle of economy of postulates leads, for a polarizable material, to additional nonuniqueness theorems; in fact, we shall derive a variety of formulas that are actually equivalent as regards any observable quantities, but that are definitely incompatible as regards formulas for the "stresses".

The possibility of deriving formulas that are seemingly incompatible, but actually equivalent, arises from the fact that there is no unique way of separating the force exerted by one part of a polarized body on an adjacent part into a long-range term, to be calculated by megascopic magnetostatics, and a short-range term, to be calculated by means of "stresses". Thus the long-range magnetic force can be calculated by use of a volume dipole density, of volume and surface pole densities, or of Amperian volume and surface current densities; the corresponding "stresses" are all different. As one must not combine volume pole densities with surface current densities, so one must not combine long-range forces computed by one method with "stresses" corresponding to another. It will turn out that with respect to observable quantities, the various methods are equivalent; however, such equivalence will not be assumed in advance. Instead, a single point of view will be adhered to thruout; and the equivalence of the others, when it holds, will be demonstrated by mathematical transformation of the formulas to alternative forms.

The point of view adopted is the following. Electrostatic and magnetostatic forces will be considered to be exerted directly by matter on matter, at a distance, not by way of stresses in the "ether"; and in a polarized medium, the element by or on which the force is exerted will be taken to be the dipole. The separation of a force into "long-range" and "short-range" terms then requires an arbitrary definition when, for instance, the body of matter τ, the force on which is to be calculated, is itself a part of a larger physical body B and is therefore in direct contact with the part τ' of B that is not included in τ. In this case a "force" can be formally calculated by integrating over τ and over τ' the formula for the magnetic force exerted by a dipole of moment $\boldsymbol{M}' d\tau'$ on a dipole of moment $\boldsymbol{M} d\tau$, with $\boldsymbol{M}$ the megascopic magnetization (magnetic moment per unit volume) and $d\tau$ the element of volume. The *long-range force*, or *magnetic force*, exerted by τ' on τ is hereafter defined as the force formally calculated by this method. The *stresses* must then include, besides short-range forces of purely mechanical origin, any short-range deviations of the actual magnetic force (as it would be calculated on the basis of an accurate atomic model) from the long-range force as just defined; such deviations must be expected because the continuous dipole density is a poor approximation to the actual distribution of polarized atoms or molecules when the elements $d\tau$ and $d\tau'$ are close together.

With this definition of long-range forces and stresses, the entire calculation could, in principle, be carried out without ever introducing magnetic pole densities or Amperian current densities, and without ever defining the megascopic field vectors $\boldsymbol{E}$, $\boldsymbol{D}$, $\boldsymbol{H}$, and $\boldsymbol{B}$ at points of a polarized medium. The formulas would, however, be cumbersome. The pole and current densities will therefore be used freely, but always merely as mathematical devices, to which no physical significance will be attributed. Similarly, the megascopic field vectors, at points of a polarized medium, will be used, but always merely as auxiliary quantities, defined mathematically; if they have any physical significance, it must be deduced from the definition. Whatever formal use may be made of pole and current densities, the terms "long-range force" and "stress" will always have the meanings explained in the last paragraph, unless the contrary is explicitly stated.

The use of the dipole, or electric or magnetic moment, as the basic element implies that a force will be considered as exerted on the matter in a volume τ if it is exerted on a charge attached to an atom with center in τ — even tho, at the instant under consideration, that charge has been displaced across the surface bounding τ. Atomic concepts, however, will not be introduced explicitly.

5.2. Sketch of the procedure. The first step (§ 5.3) is to review and put into various forms the formulas of § 2.5 for the magnetic force and torque exerted on a whole magnetized body (the "test body") by sources of field entirely outside it, and separated from it by distances large on the molecular scale. This step is easy.

The second step (§ 5.4), which requires more care, is to evaluate the long-range part of the magnetic force and torque exerted on the matter in an arbitrary volume τ of a magnetized body by all sources outside τ, including the rest of the body. In accordance with the definition of the long-range force already outlined, the procedure for calculating it is the following: First calculate, by use of continuous dipole densities, the magnetic force and torque exerted by everything outside a closed surface S_1 on everything inside a closed surface S_2 completely surrounded by S_1, and separated from it everywhere by a distance molecularly large (Fig. 12); then find the limit of this force and torque as S_1 and S_2 are brought together to form a single surface enclosing a volume τ.

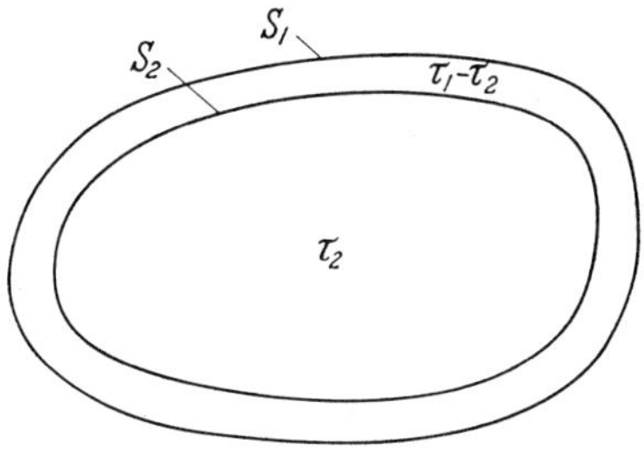

Fig. 12. Method of calculating the force on part of a body

The force and torque calculated in this manner are reliable estimates of physically significant quantities only before the limiting process is carried out; when the distance between S_1 and S_2 becomes of a molecular or atomic order of magnitude, the continuous dipole densities cease to give reliable results. The errors, however, correspond to forces that fall off with distance faster than do dipole forces; because of this short-range character, they can be taken into account, megascopically, as "stresses". In addition, there are other short-range forces, of non-magnetic origin, that act across the surface S. In a megascopic theory there is no possibility of separating these two contributions to the stress — in fact, the separation into two terms will vary with the atomic model adopted. However, regardless of the relative contributions of mechanical and magnetic terms, the resultant stress components must obey certain laws that can be deduced by megascopic theory alone. The deduction of these laws (§ 5.5), culminating in equations of (quasi-static) motion and of equilibrium for an arbitrarily magnetized and deformed body of arbitrary magnetic and mechanical properties, completes the theory of § 5. In § 6 we go on to energy and stress-strain relations.

5.3. The forces on a whole body. The magnetic field of the external sources can be described by either of two vectors, $\boldsymbol{H}_0$ and $\boldsymbol{B}_0$. The first is calculated by adding to the field intensity of the magnetizing currents, Eq. (2.4), the Coulomb field intensity of the poles in and on external magnetized bodies, Eq. (2.35). The second is calculated by adding to the field intensity of the magnetizing currents that of the Amperian currents in and on external magnetized bodies, Eq. (2.34). Since the field of the test body itself is not included in $\boldsymbol{H}_0$ or $\boldsymbol{B}_0$, the two are equal thruout the space occupied by it, and the dipole formula (2.15) (first expression) can also be used.

The magnetic moment of the matter in a volume element $d\tau$ of the test body is $\boldsymbol{M}\,d\tau$, where $\boldsymbol{M}$ is the magnetization in $d\tau$. The force exerted on it by the sources of the external field is $\boldsymbol{M}\cdot\nabla\boldsymbol{H}_0\,d\tau$, and there is also a couple $\boldsymbol{M}\times\boldsymbol{H}_0\,d\tau$ [Eqs. (2.11)], so that the torque about an arbitrary origin O — with respect to which the position vector of $d\tau$ is $\boldsymbol{r}$ — is $[\boldsymbol{r}\times(\boldsymbol{M}\cdot\nabla\boldsymbol{H}_0)+\boldsymbol{M}\times\boldsymbol{H}_0]\,d\tau$. The total force and torque are found by integrating these expressions over the volume τ occupied by the body. From the point of view adopted here, these formulas are fundamental.

Alternative expressions for the *total* force and torque on the body can be derived by use of the following vector identities:

$$\int \boldsymbol{n}\cdot\boldsymbol{u}\,\boldsymbol{v}\,dS-\int(\nabla\cdot\boldsymbol{u})\,\boldsymbol{v}\,d\tau=\int\boldsymbol{u}\cdot\nabla\boldsymbol{v}\,d\tau, \tag{5.1}$$

$$\left.\begin{aligned}&-\int(\boldsymbol{n}\times\boldsymbol{u})\times\boldsymbol{v}\,dS+\int(\nabla\times\boldsymbol{u})\times\boldsymbol{v}\,d\tau\\&\qquad=\int[\boldsymbol{u}\cdot\nabla\boldsymbol{v}+\boldsymbol{u}\times(\nabla\times\boldsymbol{v})-\boldsymbol{u}\,\nabla\cdot\boldsymbol{v}]\,d\tau,\end{aligned}\right\} \tag{5.1'}$$

$$\int\boldsymbol{n}\cdot\boldsymbol{u}\,\boldsymbol{r}\times\boldsymbol{v}\,dS-\int(\nabla\cdot\boldsymbol{u})\,\boldsymbol{r}\times\boldsymbol{v}\,d\tau=\int[\boldsymbol{r}\times(\boldsymbol{u}\cdot\nabla\boldsymbol{v})+\boldsymbol{u}\times\boldsymbol{v}]\,d\tau, \tag{5.2}$$

$$\left.\begin{aligned}&-\int\boldsymbol{r}\times[(\boldsymbol{n}\times\boldsymbol{u})\times\boldsymbol{v}]\,dS+\int\boldsymbol{r}\times[(\nabla\times\boldsymbol{u})\times\boldsymbol{v}]\,d\tau\\&\qquad=\int\{\boldsymbol{r}\times[\boldsymbol{u}\cdot\nabla\boldsymbol{v}+\boldsymbol{u}\times(\nabla\times\boldsymbol{v})-\boldsymbol{u}\,\nabla\cdot\boldsymbol{v}]+\boldsymbol{u}\times\boldsymbol{v}\}\,d\tau.\end{aligned}\right\} \tag{5.2'}$$

Here, as usual, dS is an element of the surface bounding τ, and $\boldsymbol{n}$ is the outward normal; $\boldsymbol{u}$ and $\boldsymbol{v}$ are any vector functions that satisfy the conditions for validity of the divergence theorem in the region τ. The desired formulas are obtained by setting $\boldsymbol{u}=\boldsymbol{M}$, $\boldsymbol{v}=\boldsymbol{H}_0=\boldsymbol{B}_0$. In anticipation of later extensions, the external field intensity will be written $\boldsymbol{H}_0$ when it is used with the "pole densities" $-\nabla\cdot\boldsymbol{M}$ and $\boldsymbol{n}\cdot\boldsymbol{M}$, and $\boldsymbol{B}_0$ when it is used with the "Amperian current densities" $c\nabla\times\boldsymbol{M}$ and $-c\boldsymbol{n}\times\boldsymbol{M}$; and $\nabla\times\boldsymbol{B}_0$ will be retained in the formulas (altho it actually vanishes) as a warning against forgetting $\nabla\times\boldsymbol{B}$ in later formulas. However, $\nabla\cdot\boldsymbol{B}_0$ will be omitted, since $\nabla\cdot\boldsymbol{B}=0$ always and $\nabla\cdot\boldsymbol{D}=0$ in an uncharged dielectric.

The force $\boldsymbol{F}$ and torque $\boldsymbol{L}$ are then given by

$$\boldsymbol{F}=\int\boldsymbol{M}\cdot\nabla\boldsymbol{H}_0\,d\tau \tag{5.3}$$

$$=\int[\boldsymbol{M}\cdot\nabla\boldsymbol{B}_0+\boldsymbol{M}\times(\nabla\times\boldsymbol{B}_0)]\,d\tau \tag{5.3'}$$

$$=\int\boldsymbol{n}\cdot\boldsymbol{M}\,\boldsymbol{H}_0\,dS+\int(-\nabla\cdot\boldsymbol{M})\,\boldsymbol{H}_0\,d\tau \tag{5.4}$$

$$=\int(-\boldsymbol{n}\times\boldsymbol{M})\times\boldsymbol{B}_0\,dS+\int(\nabla\times\boldsymbol{M})\times\boldsymbol{B}_0\,d\tau, \tag{5.4'}$$

$$\boldsymbol{L}=\int\boldsymbol{r}\times(\boldsymbol{M}\cdot\nabla\boldsymbol{H}_0)\,d\tau+\int\boldsymbol{M}\times\boldsymbol{H}_0\,d\tau \tag{5.5}$$

$$=\int\boldsymbol{r}\times[\boldsymbol{M}\cdot\nabla\boldsymbol{B}_0+\boldsymbol{M}\times(\nabla\times\boldsymbol{B}_0)]\,d\tau+\int\boldsymbol{M}\times\boldsymbol{B}_0\,d\tau \tag{5.5'}$$

$$=\int\boldsymbol{r}\times(\boldsymbol{n}\cdot\boldsymbol{M}\,\boldsymbol{H}_0)\,dS+\int\boldsymbol{r}\times[(-\nabla\cdot\boldsymbol{M})\,\boldsymbol{H}_0]\,d\tau \tag{5.6}$$

$$=\int\boldsymbol{r}\times[(-\boldsymbol{n}\times\boldsymbol{M})\times\boldsymbol{B}_0]\,dS+\int\boldsymbol{r}\times[(\nabla\times\boldsymbol{M})\times\boldsymbol{B}_0]\,d\tau. \tag{5.6'}$$

If there are conduction currents of volume density $\boldsymbol{J}_c$ in τ (but not crossing its surface S), there is an additional term $\int(\boldsymbol{J}_c/c)\times\boldsymbol{B}_0\,d\tau$ in the force; the corresponding term in the electric case, when there is a conduction charge density ϱ_c, is $\int\varrho_c\,\boldsymbol{E}_0\,d\tau$. Insertion of $\boldsymbol{r}\times$ gives the corresponding terms in the torques. These additional terms will not be needed in our analysis and will be omitted hereafter.

Formulas (5.4) and (5.6) would be obtained directly if magnetization were interpreted as a displacement of poles and if the force on $d\tau$ were defined as the force exerted on those poles that happened, at the instant under consideration, to be in $d\tau$. Formulas (5.4′) and (5.6′) would be obtained directly by a similar calculation based on Amperian currents. These four formulas will be treated here as having only mathematical significance, in the sense that they are derived by mathematical transformation from the formulas taken as basic: Eqs. (5.3) and (5.5), or their equivalents Eqs. (5.3′) and (5.5′).

5.4. The long-range forces on part of a body. In Fig. 12, let $\boldsymbol{H}_0=\boldsymbol{B}_0$ be the field intensity of all sources outside S_1, and let the "test body" be the part of the actual body that is inside S_2. Then the force and torque are given by the equations of § 5.3 with the integrals taken over τ_2 and S_2; but now

$$\boldsymbol{H}_0=\boldsymbol{H}-\boldsymbol{H}_{12}, \tag{5.7}$$

$$\boldsymbol{B}_0=\boldsymbol{B}-\boldsymbol{B}_{12}, \tag{5.7'}$$

where $\boldsymbol{H}$ and $\boldsymbol{B}$ are the total intensity and flux density, respectively, and $\boldsymbol{H}_{12}$ and $\boldsymbol{B}_{12}$ are the parts of them contributed by the matter in τ_1, calculated at points inside τ_2. Each term in $\boldsymbol{H}_0$ must be calculated by use of the pole formula, Eq. (2.35), and each term in $\boldsymbol{B}_0$ by use of the Amperian current formula (2.34); the dipole integral (2.15) (first expression) is now semiconvergent and can be used only if Kelvin cavities are introduced, a procedure that will be avoided here.

From this point on, the entire analysis can be carried out in two equivalent but formally different ways, corresponding to choice of $\boldsymbol{H}$ or of $\boldsymbol{B}$ as the field vector to be used. It is important to obtain both sets of results, to avoid any danger of misconstruing one or the other as evidence of the greater physical significance of the corresponding field vector. The derivation will in each case be given in detail only for the $\boldsymbol{H}$-procedure; the $\boldsymbol{B}$-procedure is generally analogous, in the "antiparallel" sense (COULOMB's Law Committee [*1*], pp. 20—23).

When $\boldsymbol{F}$ and $\boldsymbol{L}$ have been separated into two terms by use of Eq. (5.7), the limiting process S_1, $S_2 \to S$ may be carried out at once for the term involving the total intensity $\boldsymbol{H}$. Thus

$$\boldsymbol{F} = \int \boldsymbol{M} \cdot \nabla \boldsymbol{H}\, d\tau + \boldsymbol{F}_1, \tag{5.8}$$

where

$$\boldsymbol{F}_1 = -\lim_{S_1,\, S_2 \to S} \left\{ \int_{S_2} \boldsymbol{n}_2 \cdot \boldsymbol{M}_2\, \boldsymbol{H}_{12}\, dS_2 + \int_{\tau_2} (-\nabla_2 \cdot \boldsymbol{M}_2)\, \boldsymbol{H}_{12}\, d\tau_2 \right\}. \tag{5.9}$$

Now by Eq. (2.35),

$$\boldsymbol{H}_{12} = \frac{\gamma}{4\pi} \left\{ \int_{S_1} \boldsymbol{n}_1 \cdot \boldsymbol{M}_1 \frac{\mathbf{1}_{12}}{r_{12}^2}\, dS_1 + \int_{\tau_1} (-\nabla_1 \cdot \boldsymbol{M}_1) \frac{\mathbf{1}_{12}}{r_{12}^2}\, d\tau_1 \right\}, \tag{5.10}$$

where $\mathbf{1}_{12}$ is a unit vector pointing from surface or volume element 1 to surface or volume element 2, and r_{12} is the mutual distance. Insertion of Eq. (5.10) in Eq. (5.9) shows, since $\mathbf{1}_{12} = -\mathbf{1}_{21}$, that $\boldsymbol{F}_1$ can also be written

$$\boldsymbol{F}_1 = +\lim_{S_1,\, S_2 \to S} \left\{ \int_{S_1} \boldsymbol{n}_1 \cdot \boldsymbol{M}_1\, \boldsymbol{H}_{21}\, dS_1 + \int_{\tau_1} (-\nabla_1 \cdot \boldsymbol{M}_1)\, \boldsymbol{H}_{21}\, d\tau_1 \right\} \tag{5.11}$$

(*i.e.*, the Coulomb forces between the pole distributions 1 and 2 obey the law of action and reaction). If the two expressions (5.9) and (5.11) are added and divided by 2, and the limiting process is then carried out, the volume term vanishes in the limit; but the surface term does not, for $\boldsymbol{H}_{12}$ is always evaluated *inside* S_1, whereas $\boldsymbol{H}_{21}$ is always evaluated *outside* S_2. The limit is

$$\boldsymbol{F}_1 = \tfrac{1}{2} \int \boldsymbol{n} \cdot \boldsymbol{M} (\boldsymbol{H}_1^+ - \boldsymbol{H}_1^-)\, dS, \tag{5.12}$$

where $\boldsymbol{H}_1$ is the Coulomb field intensity of the poles associated with the magnetized matter in τ, and $+$ and $-$ denote the values just ouside and just inside S.

The discontinuity in $\boldsymbol{H}_1$ across the surface S affects the normal component only and is of amount $\gamma \boldsymbol{n} \cdot \boldsymbol{M} = \gamma M_n$ [see Eqs. (2.31)]. Therefore

$$\boldsymbol{H}_1^+ - \boldsymbol{H}_1^- = \boldsymbol{n} \gamma M_n, \tag{5.13}$$

and

$$\boldsymbol{F}_1 = \tfrac{1}{2} \gamma \int \boldsymbol{n} M_n^2 \, dS. \tag{5.14}$$

The total long-range force exerted on the matter in τ by everything outside it is, finally,

$$\boldsymbol{F} = \int \boldsymbol{M} \cdot \nabla \boldsymbol{H} \, d\tau + \tfrac{1}{2} \gamma \int \boldsymbol{n} \, M_n^2 \, dS. \tag{5.15}$$

The analogous $\boldsymbol{B}$-calculation is somewhat more troublesome, but the result can be obtained directly from Eq. (5.15) by setting $\boldsymbol{H} = \boldsymbol{B} - \gamma \boldsymbol{M}$:

$$\boldsymbol{F} = \int [\boldsymbol{M} \cdot \nabla \boldsymbol{B} + \boldsymbol{M} \times (\nabla \times \boldsymbol{B})] \, d\tau - \tfrac{1}{2} \gamma \int \boldsymbol{n} \, M_t^2 \, dS, \tag{5.15'}$$

where M_t is the tangential component of $\boldsymbol{M}$ on S.

A similar calculation leads to torque formulas that differ from the force formulas (5.15) and (5.15′) only by the presence of the factor $\boldsymbol{r} \times$ after each integral sign and of a couple term

$$\int \boldsymbol{M} \times \boldsymbol{H} \, d\tau = \int \boldsymbol{M} \times \boldsymbol{B} \, d\tau.$$

In the final force and torque formulas, the terms containing $\boldsymbol{H}$ or $\boldsymbol{B}$ can be transformed by use of Eqs. (5.1)—(5.2′) to the alternative form involving pole densities or Amperian current densities.

Because of the peculiar form of the last term in Eq. (5.15) or Eq. (5.15′), the long-range force cannot be expressed as simply so much per unit volume; it depends also on the shape of the volume considered.

5.5. The stresses. In the standard analysis of stresses, as reviewed in § 3.1, two assumptions are made that are not justified here. One is that any forces acting on the body, other than surface tractions, are of such nature that the force on a volume element $d\tau$ can be expressed as a force per unit volume multiplied by $d\tau$. This is not true of the long-range force just computed, which contains a shape-dependent term. The other assumption is that the torque per unit volume about a point inside $d\tau$, produced by these forces, vanishes with vanishing $d\tau$. This is not true of the magnetic forces, which produce a couple $\boldsymbol{M} \times \boldsymbol{H} \, d\tau = \boldsymbol{M} \times \boldsymbol{B} \, d\tau$ proportional to $d\tau$ (the $\boldsymbol{C} \, d\tau$ of § 3.1). Accordingly, it is not permissible to take over the standard formulas of elasticity. Instead, it is necessary to repeat the analysis with the additional force and couple terms included.

When this is done, the following results are obtained instead of Eqs. (3.2)—(3.3), (3.5)—(3.7), and (3.9):

$$\tau_1(\boldsymbol{n}) = [\tau_{11} + \tfrac{1}{2}\gamma(M_1^2 - M_{\rm n}^2)]\, n_1 + \tau_{12}\, n_2 + \tau_{13}\, n_3, \ldots, \tag{5.16}$$

$$\tau_{23} - \tau_{32} = (\boldsymbol{M} \times \boldsymbol{H})_1 = M_2 H_3 - M_3 H_2, \ldots, \tag{5.17}$$

$$\left.\begin{aligned} &\frac{\partial}{\partial x_1}\left(\tau_{11} + \frac{1}{2}\gamma M_1^2\right) + \frac{\partial \tau_{12}}{\partial x_2} + \frac{\partial \tau_{13}}{\partial x_3} + \\ &\qquad + \boldsymbol{M}\cdot\nabla \boldsymbol{H}_1 + \varrho f_1 = \varrho\, a_1, \ldots \quad \text{in} \quad V, \end{aligned}\right\} \tag{5.18}$$

$$\tau_1(\boldsymbol{n}) = T_1, \ldots \quad \text{on} \quad S, \tag{5.19}$$

$$\tau_{\alpha\beta}^* = \sum_i \sum_j l_{\alpha i}\, l_{\beta j}\, \tau_{ij} + \tfrac{1}{2}\gamma \sum_i l_{\alpha i}\, l_{\beta i}\, M_i^2 - \tfrac{1}{2}\gamma\, M_\alpha^{*2}\, \delta_{\alpha\beta}. \tag{5.20}$$

We write τ instead of t in order to reserve the symbol t for later use; $\tau_i(\boldsymbol{n})$ is the i component of force across unit area normal to $\boldsymbol{n}$, and τ_{ij} is the value of $\tau_i(\boldsymbol{n})$ when $\boldsymbol{n}$ is the unit vector along the x_j axis. The summation convention has not been used here because it is not very useful (one would have to qualify it with the phrase "not summed over ..." in several places); instead, the first four equations have been written in full for one component, with "...," to indicate other equations obtained by cyclic permutation of indices, and Eq. (5.20) has been written with all summations explicitly indicated. This last equation shows that in a rotation of axes, τ_{ij} does not have the transformation properties of a tensor (unless $\boldsymbol{M} = 0$); it is this fact that renders the summation convention unhelpful. Eq. (5.16), however, transforms to an equation of the same form in the x_α^* axes.

The same equations in terms of $\boldsymbol{B}$ rather than of $\boldsymbol{H}$ read

$$\tau_1(\boldsymbol{n}) = [\tau_{11} - \tfrac{1}{2}\gamma(M_2^2 + M_3^2 - M_{\rm t}^2)]\, n_1 + \tau_{12}\, n_2 + \tau_{13}\, n_3, \ldots, \tag{5.16'}$$

$$\tau_{23} - \tau_{32} = (\boldsymbol{M} \times \boldsymbol{B})_1 = M_2 B_3 - M_3 B_2, \ldots, \tag{5.17'}$$

$$\left.\begin{aligned} &\frac{\partial}{\partial x_1}\left[\tau_{11} - \frac{1}{2}\gamma(M_2^2 + M_3^2)\right] + \frac{\partial \tau_{12}}{\partial x_2} + \frac{\partial \tau_{13}}{\partial x_3} + \\ &\qquad + [\boldsymbol{M}\cdot\nabla \boldsymbol{B} + \boldsymbol{M} \times (\nabla \times \boldsymbol{B})]_1 + \varrho f_1 = \varrho\, a_1, \ldots \quad \text{in} \quad V. \end{aligned}\right\} \tag{5.18'}$$

Eqs. (5.19) and (5.20) are unchanged.

5.6. Alternative definitions of the "stresses". As has already been emphasized, the separation of the forces into long-range or "magnetic" forces and short-range forces or "stresses" is not unique. Instead of defining the "magnetic" force as we did in § 5.1, we might have defined it by interpreting the elements of pole strength[1] $(-\nabla\cdot\boldsymbol{M})\,d\tau$ and

[1] In this alternative model, the surface pole elements $\boldsymbol{n}\cdot\boldsymbol{M}\,dS$ exist only on the actual external bounding surface of the specimen, and not on an internal surface across which $\boldsymbol{t}(\boldsymbol{n})$ is being computed.

$\boldsymbol{n}\cdot\boldsymbol{M}\,dS$, naively and literally, as being subject to forces $(-\nabla\cdot\boldsymbol{M})\boldsymbol{H}\,d\tau$ and $(\boldsymbol{n}\cdot\boldsymbol{M})\overline{\boldsymbol{H}}\,dS$ respectively.[1] The calculation is then much simpler than that of §§ 5.2—5.3; we have only to repeat the analysis of § 3.1 with the substitutions $\varrho\boldsymbol{f}\to\varrho\boldsymbol{f}+(-\nabla\cdot\boldsymbol{M})\boldsymbol{H}$ and $\boldsymbol{T}\to\boldsymbol{T}+(\boldsymbol{n}\cdot\boldsymbol{M})\times\frac{1}{2}\{\boldsymbol{H}^{+}+\boldsymbol{H}^{-}]$. If we call the resulting stress vectors $\bar{\boldsymbol{t}}(\boldsymbol{n})$ and the stress components $\bar{t}_{ij}$,[2] we find that they satisfy the following equations:

$$\bar{t}_i(\boldsymbol{n})=\bar{t}_{ij}\,n_j, \tag{5.21}$$

$$\bar{t}_{ij}=\bar{t}_{ji}, \tag{5.22}$$

$$\bar{t}_{ij,j}+(-\nabla\cdot\boldsymbol{M})H_i+\varrho f_i=\varrho a_i \quad \text{in} \quad V, \tag{5.23}$$

$$\bar{\boldsymbol{t}}(\boldsymbol{n})-(\boldsymbol{n}\cdot\boldsymbol{M})\tfrac{1}{2}[\boldsymbol{H}^{+}+\boldsymbol{H}^{-}]=\boldsymbol{T} \quad \text{on} \quad S. \tag{5.24}$$

Here we have again used the summation convention; $\bar{t}_{ij}$ transforms as a tensor. We may question whether such a naive interpretation of poles can lead to correct equations. The fact is that we can get precisely the same equations (5.21)—(5.24) by the following formal procedure, which does not require that interpretation: in Eqs. (5.16)—(5.19), *define* a new stress vector $\bar{\boldsymbol{t}}(\boldsymbol{n})$ and new stress components $\bar{t}_{ij}$ by

$$\bar{t}_i(\boldsymbol{n})\equiv\tau_i(\boldsymbol{n})+\tfrac{1}{2}\gamma M_{\mathrm{n}}^2\,n_i+H_i\,M_{\mathrm{n}}, \tag{5.25}$$

whence

$$\left.\begin{aligned}&\bar{t}_{11}=\tau_{11}+\tfrac{1}{2}\gamma M_1^2+H_1\,M_1,\ldots,\\&\bar{t}_{23}=\bar{t}_{32}=\tau_{23}+H_2\,M_3=\tau_{32}+H_3\,M_2,\ldots,\end{aligned}\right\} \tag{5.26}$$

and substitute for the τ_{ij}'s their equivalents in terms of the $\bar{t}_{ij}$'s. The substitution gives Eqs. (5.21)—(5.24); furthermore, the same substitution in Eq. (5.20) reduces it to the tensor transformation formula $\bar{t}^{*}_{\alpha\beta}=l_{\alpha i}\,l_{\beta j}\,\tau_{ij}$. Thus the naive pole model leads to correct equations of motion and equilibrium.

From the point of view adopted in §§ 5.2—5.3, there are still some reservations about the pole-model "stresses" $\bar{t}_{ij}$. The method of definition of the τ_{ij}'s insures that the force system $\boldsymbol{\tau}(\boldsymbol{n})$ determined by them contains no long-range part, *i.e.*, no part dependent, for example, on the shape of the specimen. It then follows from Eqs. (5.25), (5.26) that the force system $\bar{\boldsymbol{t}}(\boldsymbol{n})$ *does* contain such a part: namely, the part $H_i M_{\mathrm{n}}$ of $\bar{t}_i(\boldsymbol{n})$ or the part $H_i M_j$ of the tensor $\bar{t}_{ij}$. In a uniformly magnetized ferromagnetic ellipsoid in zero applied field, with the ellipsoid axes as

[1] As is well known, the force and torque on a surface distribution can be found by assigning to each element of surface charge $\sigma\,dS$ a force $\sigma\overline{\boldsymbol{H}}\,dS$, where $\overline{\boldsymbol{H}}\equiv\frac{1}{2}[\boldsymbol{H}^{+}+\boldsymbol{H}^{-}]$ is the average of the values of $\boldsymbol{H}$ on the two sides of S. This is most easily proved by transforming the volume force $\int\varrho\boldsymbol{H}\,d\tau$ to a surface integral [cf. Eq. (2.41)], letting the volume charge shrink to a surface charge, and then shrinking the surface of integration to a double surface on the two sides of the surface distribution.

[2] The bar here does *not* imply an average or a complex conjugate.

x_i axes, $H_1 = -\gamma D_1 M_1, \ldots$, where the demagnetizing factors D_i depend on the axis ratio of the ellipsoid. Thus the "stress" vector $\boldsymbol{H} M_{\mathrm{n}}$ depends not only on local conditions but also on distant ones. As long as we bear this in mind, however, and do not derive from the $\bar{t}_{ij}$'s any results mathematically dependent on their supposed freedom from distant effects, there can be no objection to using them.

A similar treatment of the Amperian current model leads to "stresses" $\bar{t}'_{ij}$ that satisfy the relations

$$\bar{t}'_i(\boldsymbol{n}) = \bar{t}'_{ij}\, n_j, \tag{5.21'}$$

$$\bar{t}'_{ij} = \bar{t}'_{ji}, \tag{5.22'}$$

$$\bar{t}'_{ij,j} + [(\nabla \times \boldsymbol{M}) \times \boldsymbol{B}]_i + \varrho f_i = \varrho a_i \quad \text{in} \quad V, \tag{5.23'}$$

$$\bar{\boldsymbol{t}}'(\boldsymbol{n}) - (-\boldsymbol{n} \times \boldsymbol{M}) \times \tfrac{1}{2}[\boldsymbol{B}^+ + \boldsymbol{B}^-] = \boldsymbol{T} \quad \text{on} \quad S. \tag{5.24'}$$

The same relations can be derived from Eqs. (5.16)—(5.19) by substituting for the τ_{ij}'s in terms of the $\bar{t}'_{ij}$'s, defined by

$$\bar{t}'_i(\boldsymbol{n}) \equiv \tau_i(\boldsymbol{n}) - \tfrac{1}{2}\gamma M_{\mathrm{t}}^2\, n_i + \boldsymbol{M} \cdot \boldsymbol{B}\, n_i - M_i\, B_{\mathrm{n}}, \tag{5.25'}$$

whence

$$\left.\begin{aligned} \bar{t}'_{11} &= \tau_{11} - \tfrac{1}{2}\gamma (M_2^2 + M_3^2) + M_2 B_2 + M_3 B_3, \ldots, \\ \bar{t}'_{23} &= \bar{t}'_{32} = \tau_{23} - M_2 B_3 = \tau_{32} - M_3 B_2, \ldots. \end{aligned}\right\} \tag{5.26'}$$

Here, also, the "stresses" contain some terms of questionable "short-range" character but nevertheless lead to correct equations of motion and equilibrium.

Instead of relying on a pole or current model for guidance, we may deliberately choose a new set of "stresses" in such a way as to simplify Eqs. (5.16)—(5.20), and in particular to reduce them to tensor form. There are two easy ways of doing this, each of which avoids introducing any questionable "long-range" terms into the "stresses". One way is to remove the surface term $\frac{1}{2}\gamma \int \boldsymbol{n} M_{\mathrm{n}}^2\, dS$ in Eq. (5.15) from the "magnetic force" and, instead, add a term $\frac{1}{2}\gamma \boldsymbol{n} M_{\mathrm{n}}^2$ to the "stress vector". This is clearly permissible, since (1) the total force and torque on an arbitrary volume are not altered by this shift and (2) the term thus added to the stress vector is dependent only on local quantities and therefore may legitimately be considered a part of the "stress". The new stress vector is then

$$t_i(\boldsymbol{n}) = \tau_i(\boldsymbol{n}) + \tfrac{1}{2}\gamma M_{\mathrm{n}}^2\, n_i, \tag{5.27}$$

which for $\boldsymbol{n}$ along a coordinate axis becomes

$$t_{11} = \tau_{11} + \tfrac{1}{2}\gamma M_1^2, \ldots, t_{23} = \tau_{23}, \ldots, t_{32} = \tau_{32}, \ldots. \tag{5.28}$$

It may now be verified, either by substitution in Eqs. (5.16)—(5.20) or by repeating the calculation with the new definition of "magnetic" force, that

$$t_i(\boldsymbol{n}) = t_{ij}\, n_j, \tag{5.29}$$

$$t_{[ij]} = M_{[i}\, H_{j]}, \tag{5.30}$$

$$t_{ij,j} + M_j\, H_{i,j} + \varrho f_i = \varrho\, a_i \quad \text{in} \quad V, \tag{5.31}$$

$$t_{ij}\, n_j - \tfrac{1}{2}\gamma M_{\mathrm{n}}^2\, n_i = T_i \qquad \text{on} \quad S. \tag{5.32}$$

The other way is to remove the surface term $-\frac{1}{2}\gamma\int \boldsymbol{n} M_{\mathrm{t}}^2\, dS$ in Eq. (5.15′) from the "magnetic force" and, instead, add a term $-\frac{1}{2}\gamma\boldsymbol{n} M_{\mathrm{t}}^2$ to the "stress vector". The new stress vector is then

$$\left.\begin{aligned} t_i'(\boldsymbol{n}) &= \tau_i(\boldsymbol{n}) - \tfrac{1}{2}\gamma M_{\mathrm{t}}^2\, n_i \\ &= t_i(\boldsymbol{n}) - \tfrac{1}{2}\gamma \boldsymbol{M}^2\, n_i, \end{aligned}\right\} \tag{5.27′}$$

whence

$$t_{11}' = \tau_{11} - \tfrac{1}{2}\gamma(M_2^2 + M_3^2), \ldots, t_{23}' = \tau_{23}, \ldots, t_{32}' = \tau_{32}, \ldots, \tag{5.28′}$$

$$t_i'(\boldsymbol{n}) = t_{ij}\, n_j, \tag{5.29′}$$

$$t_{[ij]}' = M_{[i}\, B_{j]}, \tag{5.30′}$$

$$\left.\begin{aligned} &t_{ij,j}' + M_j\, B_{i,j} + [\boldsymbol{M} \times (\nabla \times \boldsymbol{B})]_i + \varrho f_i \\ &\quad = t_{ij,j}' + M_j\, B_{j,i} + \varrho f_i = \varrho\, a_i \quad \text{in} \quad V, \end{aligned}\right\} \tag{5.31′}$$

$$t_{ij}'\, n_j + \tfrac{1}{2}\gamma M_{\mathrm{t}}^2\, n_i = T_i \qquad \text{on} \quad S. \tag{5.32′}$$

Both t_{ij} and t_{ij}' transform as tensors, so that it was again possible to use the summation convention in Eqs. (5.29)—(5.32) and (5.29′)—(5.32′). If, however, we seek to interpret the new "magnetic" forces on element $d\tau$, $\boldsymbol{M} \cdot \nabla \boldsymbol{H}\, d\tau$ and $[\boldsymbol{M} \cdot \nabla \boldsymbol{B} + \boldsymbol{M} \times (\nabla \times \boldsymbol{B})]\, d\tau$, on the basis of a self-consistent model, we have difficulties. The first of these expressions is the force exerted on an element of magnetic moment by sources whose field is $\boldsymbol{H}$, namely the external sources plus all the poles in and on the body. The second expression is the force exerted on an element of magnetic moment by sources whose field is $\boldsymbol{B}$, namely the external sources and all the Amperian currents in and on the body. In either case the sources of the field and the objects on which it acts have not been treated consistently. That such mixed-model treatments should give correct results is somewhat surprising; but in fact they do, as our derivation shows.

Toupin's [*1*] "stresses" are our t_{ij}'s; they were derived, as was stated in § 2.5, by taking over from Maxwell a force formula to which Maxwell's two editors both raised objections. On the basis of our discussion of the formula in § 2.5 and our interpretation of it in the last paragraph, we must regard the formula as an unsatisfactory starting point for the theory. The present derivation, however, shows that TOUPIN'S final formulas are nevertheless correct.

We have now derived five sets of formulas, corresponding to five different separations of a total force into a "magnetic" and a "stress" part. In every respect except mathematical convenience, the original "stresses" τ_{ij} are preferable: they are based on a self-consistent definition of "magnetic force" and include as "stresses" only forces that are genuinely of short-range character. In mathematical convenience, however, they are inferior to all the others.

To resolve this dilemma, we shall adhere to our original definitions, so that the "magnetic force" includes the surface terms in Eqs. (5.15) and (5.15′) and the "stress" vector is $\tau_i(\boldsymbol{n})$; but we shall, for ease of mathematical manipulation, work chiefly with the quantities $t_i(\boldsymbol{n})$ and t_{ij}, which we shall regard strictly as auxiliary quantities defined by Eqs. (5.27), (5.28).

6. Equilibrium in a Magnetizable Elastic Solid

6.1. Energy relations. In §2.6 we calculated the work done by the battery in the magnetizing circuit when a rigid magnetic body undergoes a change of magnetization. In § 3.3 we calculated the work done by the externally applied body and surface forces when a nonmagnetic elastic body undergoes a change of deformation. We now wish to modify these calculations so that both apply to a body that is both magnetizable and deformable. We shall then be able to calculate the total work done by both sets of force.

In this calculation, one can attempt to calculate the work done by the forces that act on an arbitrary part of a body, or one can be content to calculate the work done on the whole body. The former procedure requires certain postulates that are not required by the latter: namely, postulates about the localization of the internal forces. An example of this occurred in § 3.3, where two different sets of postulates about the forces $\boldsymbol{t}(\boldsymbol{n})$, a stronger and a weaker, led to two sets of conclusions, a stronger and a weaker. In this section we shall calculate the work relations for *an arbitrary part* of a magnetized body; and in doing so, we shall adopt the same definitions as in § 5.4: that is, we shall regard magnetic forces as forces exerted by magnetic moment elements $\boldsymbol{M}_1\, d\tau_1$ on magnetic moment elements $\boldsymbol{M}_2\, d\tau_2$ [rather than by pole elements $(-\nabla_1 \cdot \boldsymbol{M}_1)\, d\tau_1$ on pole elements $(-\nabla_2 \cdot \boldsymbol{M}_2)\, d\tau_2$, or by Amperian current elements $c(\nabla_1 \times \boldsymbol{M}_1)\, d\tau_1$ on Amperian current elements $c(\nabla_2 \times \boldsymbol{M}_2)\, d\tau_2$]. We shall also regard the "stress vector" $\boldsymbol{\tau}(\boldsymbol{n})$ as describing a short-range force (additional to any included in the "magnetic force") $\boldsymbol{\tau}(\boldsymbol{n})\, dS$ exerted across an arbitrary internal element of area $\boldsymbol{n}\, dS$. After we have drawn conclusions from this analysis, we shall compare them with the conclusions that could be drawn if only the work relation for *the whole body* had been calculated.

We calculate first the work by the battery against electromotive forces induced by changes (of magnetization or of deformation) in the mass m_1 that instantaneously occupies some arbitrary volume τ_1 in the body. Let Φ_{m1} be the part of the flux thru the magnetizing coil that is due to the magnetic moments of this particular mass m_1. Then from the first step in Eq. (2.46),

$$\left.\begin{aligned}\frac{\delta W_{\text{bat}}}{\delta t} &= \frac{I}{c}\frac{d\Phi_{m1}}{dt} = \frac{1}{c}\left\{\frac{d}{dt}(I\Phi_{m1}) - \Phi_{m1}\frac{dI}{dt}\right\}\\ &= \frac{d}{dt}\int\limits_{\tau_1}\boldsymbol{M}\cdot\boldsymbol{H}_{01}\,d\tau - \int\limits_{\tau_1}\boldsymbol{M}\cdot\frac{\partial \boldsymbol{H}_{01}}{\partial t}\,d\tau,\end{aligned}\right\} \tag{6.1}$$

by transformations similar to those of Eq. (2.48). Here $\partial\boldsymbol{H}_{01}/\partial t$ is the rate of change of the coil field $\boldsymbol{H}_{01}$ at a fixed point of space. It is important to notice that $\boldsymbol{H}_{01}$ has no discontinuity across the surface of the specimen or of the mass m_1 under consideration; therefore $\partial\boldsymbol{H}_{01}/\partial t$ remains finite as, in the course of specimen deformation, a surface element of the body or of m_1 moves past the point where $\partial\boldsymbol{H}_{01}/\partial t$ is being evaluated. This is not true of $\boldsymbol{H}$, the total magnetizing force, at the surface of the specimen; nor is it true of $\boldsymbol{H}_1$, the contribution to $\boldsymbol{H}$ from the moments of m_1, at the surface of m_1.[1]

We now go over to an integration over mass elements $dm = \varrho\,dV$, with moment per unit mass $\mathsf{M} = \boldsymbol{M}/\varrho$. If $d\boldsymbol{H}_{01}/dt$ is the rate of change of the applied field acting on a particular particle,

$$\frac{d\boldsymbol{H}_{01}}{dt} = \frac{\partial \boldsymbol{H}_{01}}{\partial t} + \boldsymbol{v}\cdot\nabla\boldsymbol{H}_{01}, \tag{6.2}$$

where $\boldsymbol{v}$ is the velocity of the particle. Hence

$$\left.\begin{aligned}\frac{\delta W_{\text{bat}}}{\delta t} &= \frac{d}{dt}\int\mathsf{M}\cdot\boldsymbol{H}_{01}\,dm - \int\mathsf{M}\cdot\left(\frac{d\boldsymbol{H}_{01}}{dt} - \boldsymbol{v}\cdot\nabla\boldsymbol{H}_{01}\right)dm\\ &= \int\boldsymbol{H}_{01}\cdot\frac{d\mathsf{M}}{dt}\,dm + \int\mathsf{M}\cdot(\boldsymbol{v}\cdot\nabla\boldsymbol{H}_{01})\,dm.\end{aligned}\right\} \tag{6.3}$$

If, as we shall assume, the magnetizing currents do not traverse the specimen, then $\nabla\times\boldsymbol{H}_{01} = 0$ in the specimen. It then follows that M

[1] To calculate $\boldsymbol{H}_1$, we calculate the pole-field of moments in the internal region in Fig. 12 (there called τ_2) and find the limit of this as S_1 and S_2 come together. The pole-field includes a contribution from surface poles $\boldsymbol{n}\cdot\boldsymbol{M}\,dS$ on S_2 of Fig. 12. The pole-field of the opposite poles $-\boldsymbol{n}\cdot\boldsymbol{M}\,dS$ on S_1 cancels this contribution, in the limit, in $\boldsymbol{H}$ but not in $\boldsymbol{H}_1$. In other words, the surface that bounds m_1 separates dipoles from dipoles, not charges from charges.

and $\boldsymbol{v}$ may be interchanged in the last integral: for

$$\left.\begin{aligned}\mathbf{M}\cdot(\boldsymbol{v}\cdot\nabla\boldsymbol{H}_{01}) &= \mathsf{M}_i(\boldsymbol{v}\cdot\nabla H_{01i}) = \mathsf{M}_i v_j H_{01i,j}\\ &= \mathsf{M}_i v_j H_{01j,i} = \boldsymbol{v}\cdot(\mathbf{M}\cdot\nabla\boldsymbol{H}_{01}).\end{aligned}\right\} \tag{6.4}$$

Thus, finally,

$$\frac{\delta W_{\mathrm{bat}}}{\delta t} = \int \boldsymbol{H}_{01}\cdot\frac{d\mathbf{M}}{dt}\,dm + \int \boldsymbol{v}\cdot(\mathbf{M}\cdot\nabla\boldsymbol{H}_{01})\,dm. \tag{6.5}$$

The second term is clearly the rate of work by the magnetic forces $\mathbf{M}\cdot\nabla\boldsymbol{H}_{01}\,dm$ exerted by the current I on the particles dm [cf. Eq. (2.10)].

We calculate second the rate of work by the magnetic forces exerted on m_1 by the rest of the magnetized body m_2. The mutual forces between two small magnets with moments $\boldsymbol{m}_1$ and $\boldsymbol{m}_2$, at positions $\boldsymbol{r}_1$ and $\boldsymbol{r}_2$, can be found by differentiation of an effective potential energy

$$\left.\begin{aligned}U &= \frac{\gamma}{4\pi}\,\frac{\boldsymbol{m}_1\cdot\boldsymbol{m}_2 - 3\,\boldsymbol{m}_1\cdot\mathbf{1}_{12}\,\boldsymbol{m}_2\cdot\mathbf{1}_{12}}{r_{12}^3}\\ &= -\boldsymbol{H}_{12}\cdot\boldsymbol{m}_2 = -\boldsymbol{H}_{21}\cdot\boldsymbol{m}_1,\end{aligned}\right\} \tag{6.6}$$

where $\boldsymbol{H}_{pq}$ is the field intensity of $\boldsymbol{m}_p$ at the position of $\boldsymbol{m}_q$; it is easily verified that this leads to the force and couple formulas (2.11). The work done in a change of $\boldsymbol{r}_1$, $\boldsymbol{r}_2$, $\boldsymbol{m}_1$, and $\boldsymbol{m}_2$ is therefore equal to the decrease of U; the part containing $\delta\boldsymbol{r}_1$ and $\delta\boldsymbol{m}_1$ is done by the forces exerted by $\boldsymbol{m}_2$ on $\boldsymbol{m}_1$, and the part containing $\delta\boldsymbol{r}_2$ and $\delta\boldsymbol{m}_2$ is done by the forces exerted by $\boldsymbol{m}_1$ on $\boldsymbol{m}_2$. Here we are interested in the former, namely

$$\delta W_{21} = -\delta_1 U = \boldsymbol{H}_{21}\cdot\delta\boldsymbol{m}_1 + \boldsymbol{m}_1\cdot(\delta\boldsymbol{r}_1\cdot\nabla_1\boldsymbol{H}_{21}). \tag{6.7}$$

On replacing $\boldsymbol{m}_1$ by an element $\mathbf{M}\,dm$, we get[1]

$$\frac{\delta W_{21}}{\delta t} = \int_{\tau_1} \boldsymbol{H}_{02}\cdot\frac{d\mathbf{M}}{dt}\,dm + \int_{\tau_1} \boldsymbol{v}\cdot(\mathbf{M}\cdot\nabla\boldsymbol{H}_{02})\,dm, \tag{6.8}$$

where $\boldsymbol{H}_{02}$ is the field intensity of the moments in m_2. Again we have invoked Eq. (6.4), this time applied to $\boldsymbol{H}_{02}$.

From Eqs. (6.5) and (6.8) we get for the total rate of *magnetic* work on m_1, defined as the work by batteries in the magnetizing coil plus the work by forces due to other magnetized matter,

$$\frac{\delta W_{\mathrm{mag}}}{\delta t} = \int_{\tau_1} \boldsymbol{H}_0\cdot\frac{d\mathbf{M}}{dt}\,dm + \int_{\tau_1} \boldsymbol{v}\cdot(\mathbf{M}\cdot\nabla\boldsymbol{H}_0)\,dm, \tag{6.9}$$

where $\boldsymbol{H}_0$ is the combined field intensity of the coil and of magnetized matter outside τ_1.

[1] The method used here avoids the problem of expressing $d\mathbf{M}$ in terms of more specific internal coordinates, in order to evaluate the work done by the external forces that tend to increase these coordinates. For explicit treatment of these internal coordinates for three different models, see Brown [*10*], pp. 64–66.

We now wish, as in § 2.6, to set $\boldsymbol{H}_0 = \boldsymbol{H} - \boldsymbol{H}_1$ (we forego the alternative $\boldsymbol{B} - \boldsymbol{B}_1$) and to subtract out the rate of change of the "magnetostatic self-energy" W_{m} of the mass m_1. Since the calculation of dW_{m}/dt is rather tedious, we shall in this section merely state and use the result, deferring the proof to § 6.2:

$$\left.\begin{aligned}\frac{dW_{\mathrm{m}}}{dt} = &-\int \boldsymbol{H}_1 \cdot \frac{d\mathbf{M}}{dt}\, dm - \int \boldsymbol{v} \cdot (\mathbf{M} \cdot \nabla \boldsymbol{H}_1)\, dm - \\ &-\frac{1}{2}\gamma \int \boldsymbol{v} \cdot \boldsymbol{n} M_{\mathrm{n}}^2\, dS.\end{aligned}\right\} \tag{6.10}$$

On subtracting this from $\delta W_{\mathrm{mag}}/\delta t$, we get

$$\left.\begin{aligned}\frac{\delta W_{\mathrm{mag}}}{\delta t} - \frac{dW_{\mathrm{m}}}{dt} = &\int \boldsymbol{H} \cdot \frac{d\mathbf{M}}{dt}\, dm + \int \boldsymbol{v} \cdot (\mathbf{M} \cdot \nabla \boldsymbol{H})\, dm + \\ &+\frac{1}{2}\gamma \int \boldsymbol{v} \cdot \boldsymbol{n} M_{\mathrm{n}}^2\, dS.\end{aligned}\right\} \tag{6.11}$$

The rate of work by the mechanical body forces $\boldsymbol{f}\, d\tau$ in τ_1 and by the forces $\boldsymbol{\tau}(\boldsymbol{n})$ that act across its surface is, by Eqs. (5.27)—(5.31),

$$\left.\begin{aligned}\frac{\delta W_{\mathrm{mech}}}{\delta t} &= \int \varrho \boldsymbol{f} \cdot \boldsymbol{v}\, d\tau + \int \boldsymbol{\tau}(\boldsymbol{n}) \cdot \boldsymbol{v}\, dS \\ &= \int \varrho f_i v_i\, d\tau + \int t_i(\boldsymbol{n}) v_i\, dS - \frac{1}{2}\gamma \int M_{\mathrm{n}}^2 n_i v_i\, dS \\ &= \int \{\varrho a_i - t_{ij,j} - M_j H_{i,j}\} v_i\, d\tau + \int (t_{ij} n_j - \tfrac{1}{2}\gamma M_{\mathrm{n}}^2 n_i) v_i\, dS \\ &= \int \left(\frac{dv_i}{dt}\right) v_i\, dm + \int [(-t_{ij,j} - M_j H_{i,j}) v_i + (t_{ij} v_j)_{,j}]\, d\tau - \\ &\quad - \frac{1}{2}\gamma \int M_{\mathrm{n}}^2\, \boldsymbol{n} \cdot \boldsymbol{v}\, dS \\ &= \frac{d\mathcal{T}}{dt} + \int [-\boldsymbol{v} \cdot (\boldsymbol{M} \cdot \nabla \boldsymbol{H}) + t_{ij} v_{i,j}]\, d\tau - \frac{1}{2}\gamma \int \boldsymbol{v} \cdot \boldsymbol{n} M_{\mathrm{n}}^2\, dS,\end{aligned}\right\} \tag{6.12}$$

where $\mathcal{T} = \int \frac{1}{2} \boldsymbol{v}^2\, dm$ is the kinetic energy. On adding Eqs. (6.11) and (6.12), we get for the total work $\delta W (= \delta W_{\mathrm{bat}} + \delta W_{\mathrm{mech}})$ [1]

$$\left.\begin{aligned}\frac{\delta W}{\delta t} - \frac{dW_{\mathrm{m}}}{dt} - \frac{d\mathcal{T}}{dt} &= \int \boldsymbol{H} \cdot \frac{d\mathbf{M}}{dt}\, dm + \int t_{ij} v_{i,j}\, d\tau \\ &= \int \left[H_i \frac{d\mathrm{M}_i}{dt} \varrho + t_{ij} v_{i,j}\right] d\tau.\end{aligned}\right\} \tag{6.13}$$

[1] Eq. (6.13) is equivalent to Eq. (4.2—10) of Brown [5]. The latter equation, however, was expressed in terms of an "angular velocity" $\boldsymbol{\Omega} = \frac{1}{2} \nabla \times \boldsymbol{v}$ that has no simple physical significance except within the approximations of linear elasticity theory. To put Eq. (6.13) into the form (4.2—10) we note that $t_{ij} v_{i,j} = [t_{(ij)} + t_{[ij]}]\, v_{i,j} = t_{(ij)} v_{(i,j)} + t_{[ij]} v_{[i,j]} = t_{(ij)} v_{(i,j)} + M_{[i} H_{j]} v_{[i,j]}$. The second term is equivalent to $-\frac{1}{2} \boldsymbol{M} \times \boldsymbol{H} \cdot (\nabla \times \boldsymbol{v}) = -\varrho \mathbf{M} \times \boldsymbol{H} \cdot \boldsymbol{\Omega} = -\varrho \boldsymbol{H} \cdot \boldsymbol{\Omega} \times \boldsymbol{M}$; combined with the term $H_i (d\mathrm{M}_i/dt)\varrho = \varrho \boldsymbol{H} \cdot d\mathbf{M}/dt$ in Eq. (6.13), this gives $\varrho \boldsymbol{H} \cdot D\mathbf{M}/dt$ and hence the term $\int \boldsymbol{H} \cdot (D\mathbf{M}/dt)\, dm$ in Eq. (4.2—10). The symmetric part $t_{(ij)} v_{(i,j)}$ gives the terms in ε_{xx} etc. in Eq. (4.2—10). The limitations of the "angular velocity" interpretation of $\boldsymbol{\Omega}$ are discussed and illustrated in Appendix C.

Replacement of actual by virtual variations gives

$$\delta W - \delta W_{\mathrm{m}} - \delta \mathscr{T} = [\varrho H_i \, \delta \mathsf{M}_i + t_{ij} \, \delta x_{i,j}] \, d\tau. \tag{6.14}$$

In the analogous electrostatic calculation there is no $\boldsymbol{E}_{01}$, and the sources of $\boldsymbol{E}_{02}$ may be isolated charges as well as dipoles. The formulas are the same except for substitution of electric for magnetic quantities.

6.2. The rate of change of the magnetic self-energy. A derivation of Eq. (6.6) was given in Appendix A of BROWN [5]. A derivation by finite-strain theory is given in Appendix A of this monograph. The following shorter derivation will suffice for our immediate purposes.

We seek a formula for the rate of change of $W_{\mathrm{m}} \equiv -\frac{1}{2} \int\limits_{m_1} \mathbf{M} \cdot \boldsymbol{H}_1 \, dm = \frac{1}{2\gamma} \int \boldsymbol{H}_1^2 \, d\tau$, where m_1 is a specified part of the magnetic body, instantaneously occupying a volume τ_1. As the specimen deforms, m_1 continues to contain the same particles but may occupy a different volume τ_1; $\boldsymbol{H}_1$ is the part of the magnetizing force due to all the moment elements $\boldsymbol{M} \, d\tau = \mathbf{M} \, dm$ in m_1. Since matter outside m_1 contributes nothing to W_{m}, we may imagine it removed and treat m_1 as the only matter present. Hence $\boldsymbol{H}_1$ has a discontinuity of amount $\gamma \boldsymbol{n} \cdot \boldsymbol{M} \boldsymbol{n}$ from inside to outside the surface S_1 that bounds m_1.

First, we replace the surface S_1, where the $\boldsymbol{M}$ under consideration drops discontinuously to zero, by a continuous transition. Then from $W_{\mathrm{m}} = -\frac{1}{2} \int \boldsymbol{M} \cdot \boldsymbol{H}_1 \, d\tau$ and the reciprocity relation used before, we get

$$\left.\begin{aligned} \frac{dW_{\mathrm{m}}}{dt} &= -\frac{1}{2} \int \left(\boldsymbol{M} \cdot \frac{\partial \boldsymbol{H}_1}{\partial t} + \boldsymbol{H}_1 \cdot \frac{\partial \boldsymbol{M}}{\partial t} \right) d\tau \\ &= -\int \boldsymbol{M} \cdot \frac{\partial \boldsymbol{H}_1}{\partial t} \, d\tau = -\int \mathbf{M} \cdot \left(\frac{d\boldsymbol{H}_1}{dt} - \boldsymbol{v} \cdot \nabla \boldsymbol{H}_1 \right) dm \\ &= -\int \mathbf{M} \cdot \frac{d\boldsymbol{H}_1}{dt} \, dm + \int \boldsymbol{v} \cdot (\mathbf{M} \cdot \nabla \boldsymbol{H}_1) \, dm; \end{aligned}\right\} \tag{6.15}$$

from $W_{\mathrm{m}} = -\frac{1}{2} \int \mathbf{M} \cdot \boldsymbol{H}_1 \, dm$, we get

$$\frac{dW_{\mathrm{m}}}{dt} = -\frac{1}{2} \int \mathbf{M} \cdot \frac{d\boldsymbol{H}_1}{dt} \, dm - \frac{1}{2} \int \boldsymbol{H}_1 \cdot \frac{d\mathbf{M}}{dt} \, dm; \tag{6.16}$$

on subtracting the first of these from twice the second, we get

$$\frac{dW_{\mathrm{m}}}{dt} = -\int \boldsymbol{H}_1 \cdot \frac{d\mathbf{M}}{dt} \, dm - \int \boldsymbol{v} \cdot (\mathbf{M} \cdot \nabla \boldsymbol{H}_1) \, dm. \tag{6.17}$$

These are the first two terms of Eq. (6.10). The third term is due to the surface discontinuities in M_{n} and H_{n}, which were excluded by hypothesis in the derivation just completed. To get this term, we separate out the contribution from the region $\Delta\tau$ between a surface S^- just inside the specimen surface and a surface S^+ just outside, and find the limiting value of this as the surfaces come together and the continuous transition becomes a discontinuity.

In this process, volume integrands that remain finite contribute nothing. The volume integrands in Eq. (6.17) are $H_{1i}(dM_i/dt)\varrho$ and $v_i M_j H_{1i,j}$. The first remains finite in the limit; the second does not, since $\boldsymbol{H}_1$ has a discontinuity $\gamma M_n \boldsymbol{n}$ from inside to outside. The second integrand can be rewritten[1]

$$\left.\begin{aligned} v_i M_j H_{1i,j} = \gamma^{-1}\{(v_i B_{ij} H_{1i})_{,j} - v_{i,j} B_{1j} H_{1i} \\ -\tfrac{1}{2}(v_i H_{1j} H_{1j})_{,i} + \tfrac{1}{2} v_{i,i} H_{1j} H_{1j}\}. \end{aligned}\right\} \tag{6.18}$$

Transformation to a surface integral over S^+ and S^-, followed by the limiting process described, gives

$$\left.\begin{aligned} \int_{\Delta\tau} v_i M_j H_{1i,j}\, d\tau &= \gamma^{-1} \int_{S^+-S^-} \{v_i B_{1j} H_{1i} n_j - \tfrac{1}{2} v_i H_{1j} H_{1j} n_i\}\, dS \\ &= \gamma^{-1} \int_S \{\boldsymbol{v}\cdot(\boldsymbol{H}_1^+ - \boldsymbol{H}_1^-) B_{1n} - \tfrac{1}{2} v_n (\boldsymbol{H}^{+2} - \boldsymbol{H}^{-2})\}\, dS. \end{aligned}\right\} \tag{6.19}$$

But

$$\boldsymbol{H}_1^+ - \boldsymbol{H}_1^- = \gamma \boldsymbol{n} M_n, \quad \text{and} \quad \boldsymbol{H}_1^{+2} - \boldsymbol{H}_1^{-2} = (\boldsymbol{H}^+ + \boldsymbol{H}^-)\cdot \gamma \boldsymbol{n} M_n$$
$$= 2\gamma \overline{H}_n M_n, \quad \text{where} \quad \overline{H}_n \equiv \tfrac{1}{2}(H_n^+ + H_n^-).$$

Therefore

$$\begin{aligned} \int_{\Delta\tau} v_i M_j H_{1i,j}\, dV &= \int_S \{v_n M_n B_{1n} - v_n M_n \overline{H}_n\}\, dS \\ &= \gamma \int_S v_n M_n \overline{M}_n\, dS \\ &= \tfrac{1}{2}\gamma \int_S v_n M_n^2\, dS \end{aligned} \tag{6.20}$$

---

[1] The motivation of this transformation is as follows. We wish to eliminate derivatives of $\boldsymbol{H}$, leaving only integrands that remain finite in the limit (and therefore contribute nothing) and integrands $f_{,k}$ that are themselves derivatives (so that they can be transformed to surface integrals over S^+ and S^-). We may suppose that in the limit, the external region is occupied by a nonmagnetic material whose velocity is continuous with that of the internal material; then derivatives of $\boldsymbol{v}$ remain finite in the limit. Thus if we can express the factor $M_j H_{1i,j}$ in the desired form, an integration by parts $[v_i f_{i,k} = (v_i f_i)_{,k} - v_{i,k} f_i]$ will reduce the whole integrand to that form. Now

$$\begin{aligned} M_j H_{1i,j} &= (M_j H_{1i})_{,j} - M_{j,j} H_{1i} \\ &= (M_j H_{1i})_{,j} + \gamma^{-1} H_{1j,j} H_{1i} \\ &= (M_j H_{1i})_{,j} + \gamma^{-1}(H_{1j} H_{1i})_{,j} - \gamma^{-1} H_{1j} H_{1i,j} \\ &= \gamma^{-1}(B_{1j} H_{1i})_{,j} - \gamma^{-1} H_{1j} H_{1j,i} \\ &= \gamma^{-1}[(B_{1j} H_{1i})_{,j} - \tfrac{1}{2}(H_{1j} H_{1j})_{,i}]. \end{aligned}$$

Multiplication by v_i (with the resulting implied summation over i) and integration by parts gives Eq. (6.18).

$\left(\text{since } \overline{M}_n = \frac{1}{2}(M_n + 0) = \frac{1}{2} M_n\right)$. The negative of this is the third term in Eq. (6.10).

6.3. Relations of magnetizing force and stress to magnetization and deformation. We now suppose that the work (6.13) not accounted for as magnetostatic self-energy W_m or as kinetic energy $\mathcal{T}$ is stored in local form. Let F be the local (free) energy per unit mass, and suppose that it is a function of the components M_i of the moment per unit mass and of the deformation gradients $x_{i,A}$; we are for the present ignoring exchange forces, so that we include no dependence of F on gradients (either $\mathsf{M}_{i,j}$ or $\mathsf{M}_{i,A}$) of $\mathbf{M}$ and impose no constraint on $\mathbf{M}^2$. Then the total local free energy is

$$F = \int \mathsf{F}(\mathsf{M}_i, x_{i,A})\, dm; \tag{6.21}$$

and in a virtual variation,

$$\left.\begin{aligned} \delta F &= \int \left\{ \frac{\partial \mathsf{F}}{\partial \mathsf{M}_i} \delta \mathsf{M}_i + \frac{\partial \mathsf{F}}{\partial x_{i,A}} \delta x_{i,A} \right\} dm \\ &= \int \varrho \left\{ \frac{\partial \mathsf{F}}{\partial \mathsf{M}_i} \delta \mathsf{M}_i + \frac{\partial \mathsf{F}}{\partial x_{i,A}} \delta x_{i,j}\, x_{j,A} \right\} d\tau . \end{aligned}\right\} \tag{6.22}$$

On equating this to the right member of Eq. (6.13), we get

$$\int \left\{ \varrho \left(H_i - \frac{\partial \mathsf{F}}{\partial \mathsf{M}_i} \right) \delta \mathsf{M}_i + \left(t_{ij} - \varrho \frac{\partial \mathsf{F}}{\partial x_{i,A}} x_{j,A} \right) \delta x_{i,j} \right\} d\tau = 0. \tag{6.23}$$

Since this must hold when the variations $\delta \mathsf{M}_i$ are arbitrary functions of the deformed coordinates x_i (or the undeformed coordinates X_A), we may conclude that

$$H_i = \frac{\partial \mathsf{F}}{\partial \mathsf{M}_i}. \tag{6.24}$$

This relates the components of magnetizing force $\boldsymbol{H}$ to the components of moment per unit mass $\mathbf{M}$ and to the deformation gradients $x_{i,A}$.

Furthermore, we have established (6.23) for the mass occupying *an arbitrary volume* τ_1, therefore the integrand must vanish. And since the nine quantities $\delta x_{i,j}$ *at a specified point* can be chosen arbitrarily, we may conclude that $t_{ij} - \varrho(\partial \mathsf{F}/\partial x_{i,A})\, x_{j,A}$ must vanish for each i and j. Thus

$$t_{ij} = \varrho(\partial \mathsf{F}/\partial x_{i,A})\, x_{j,A}. \tag{6.25}$$

This is the same as Eq. (3.42), with the exception that F now depends not only on the $x_{i,A}$'s but also on the M_i's — which, however, are held constant in the differentiations in Eq. (6.25).

The result (6.25) is dependent on the validity of Eq. (6.23) for an arbitrary region, and not just for the body as a whole. It is therefore dependent on our postulate that the internal force $\tau(\boldsymbol{n})\, dS$ actually acts across dS (rather than merely giving the right total force and torque on a closed internal surface S). Without this postulate, we can derive Eq. (6.23) only for the whole body. Then, as in § 3.3, we can show that t_{ij} may differ from the right member of Eq. (6.25) by a tensor P_{ij} such that $P_{ij,j}=0$ in the body and $P_{ij}n_j=0$ on its surface.[1] From the discussion in § 3.3 it appears that such a P_{ij} has no effect on observable quantities; our postulate is therefore not capable of experimental verification or refutation and must be classed as a "superfluous" postulate (cf. § 1.4). This does not mean that it is necessarily desirable or even possible to dispense with such a postulate; but it does mean that if some other theory gives "stresses" different from those derived here, the two theories are not necessarily mutually contradictory in any physical sense. The two theories may legitimately differ for two reasons: they may separate the total force differently into "magnetic" and "stress" parts, as was shown in § 5.6; and for a given separation, they may involve stresses that differ by terms of the form P_{ij} just discussed.

With these reservations, we may accept Eq. (6.25) and proceed. We now invoke the physically obvious principle that if a mass element dm is rotated rigidly, together with its magnetic moment $\mathbf{M}\,dm$, its internal (free) energy $\mathsf{F}\,dm$ should not change. Therefore if we change from the twelve variables M_i ($i=1, 2, 3$) and $x_{i,A}$ ($i, A=1, 2, 3$) to twelve other independent variables of which three are the parameters (*e.g.* Euler angles) equivalent to the rotation tensor R_{Ai}, F must be independent of these last three. A physically obvious choice for the other nine variables is the three components of $\mathbf{M}$ in axes $d\overline{X}_A=R_{Ai}\,dx_i$ that share the rotation of the mass element, and the six independent components of the strain tensor E_{AB} (or of the tensor $C_{AB}=\delta_{AB}+2E_{AB}=x_{i,A}\,x_{i,B}$). The components of $\mathbf{M}$ in the $d\overline{X}_A$ axes are

$$\overline{\mathsf{M}}_A=R_{Ai}\,\mathsf{M}_i; \tag{6.26}$$

conversely,

$$\mathsf{M}_i=R_{Ai}\,\overline{\mathsf{M}}_A. \tag{6.27}$$

We therefore set

$$\mathsf{F}(\mathsf{M}_i, x_{i,A})=\mathscr{F}(\overline{\mathsf{M}}_A, C_{PQ}). \tag{6.28}$$

Then Eq. (6.24) becomes

$$H_i=\partial\mathsf{F}/\partial\mathsf{M}_i=(\partial\mathscr{F}/\partial\overline{\mathsf{M}}_A)\,R_{Ai}, \tag{6.29}$$

[1] When we later impose the condition that $\mathsf{F}\,dm$ must be invariant with respect to a rigid rotation of dm and of its magnetic moment, we shall find that t_{ij} as given by Eq. (6.25) satisfies the asymmetry condition (5.30); therefore if the term P_{ij} is allowed, it will have to satisfy the additional condition $P_{[ij]}=0$. If P_{ij} satisfies the two conditions $P_{ij,j}=0$ and $P_{[ij]}=0$ thruout a volume τ, then the surface forces $P_{ij}n_j\,dS$ on the surface S of τ produce zero force ($\int P_{ij}n_j\,dS=\int P_{ij,j}\,d\tau=0$) and zero torque ($\int x_{[k}P_{i]j}n_j\,dS=\int(x_{[k}P_{i]j})_{,j}\,d\tau=\int(\delta_{j,[k}P_{i]j}+x_{[k}P_{i]j,j})\,d\tau=\int P_{[ik]}\,d\tau=0$).

whence

$$\overline{H}_B = R_{Bi} H_i = (\partial \mathscr{F}/\partial \overline{\mathsf{M}}_A)\,\delta_{AB} = \partial \mathscr{F}/\partial \overline{\mathsf{M}}_A, \tag{6.30}$$

since $R_{Bi} R_{Ai} = \delta_{AB}$ in consequence of the orthogonality of R. That is, differentiation of F with respect to the components of $\mathbf{M}$ in the rotating axes gives the components of $\boldsymbol{H}$ in the same axes. (These relations will have to be modified when exchange forces are introduced.)

The transformation of Eq. (6.25) is less simple. We have

$$\frac{\partial \mathsf{F}}{\partial x_{i,A}} = \frac{\partial \mathscr{F}}{\partial \overline{\mathsf{M}}_B} \mathsf{M}_j \frac{\partial R_{Bj}}{\partial x_{i,A}} + \frac{\partial \mathscr{F}}{\partial C_{PQ}} \frac{\partial C_{PQ}}{\partial x_{i,A}}, \tag{6.31}$$

in which $\partial R_{Bj}/\partial x_{i,A}$ and $\partial C_{PQ}/\partial x_{i,A}$ may be evaluated by use of Eqs. (3.17) and (3.12). After considerable algebra, Eq. (6.25) can be reduced to

$$\left.\begin{aligned} t_{ik} = \varrho x_{k,A} \frac{\partial \mathsf{F}}{\partial \overline{\mathsf{M}}_B} \mathsf{M}_i (C^{-\frac{1}{2}})_{BA} + \varrho \frac{\partial \mathsf{F}}{\partial \overline{\mathsf{M}}_B} \mathsf{M}_j x_{k,A} x_{j,C} \frac{\partial (C^{-\frac{1}{2}})_{BC}}{\partial x_{i,A}} + \\ + \varrho \left(\frac{\partial \mathscr{F}}{\partial C_{AP}} + \frac{\partial \mathscr{F}}{\partial C_{PA}} \right) x_{k,A} x_{i,P}. \end{aligned}\right\} \tag{6.32}$$

The first term can be reduced to $M_i H_k$ and constitutes the antisymmetric part of t_{ik}, in accordance with Eq. (5.30); the other terms can be shown to be symmetric. Further consideration of relations for a completely general function $\mathscr{F}$ seems unprofitable; it is better to insert the particular function $\mathscr{F}$ (such as a power series truncated after a few terms) directly in Eq. (6.25) and to carry out the differentiations and the algebra for the particular case. The main difficulty in any general calculation based on Eq. (6.32) is the evaluation of the derivatives $\partial (C^{-\frac{1}{2}})_{BC}/\partial x_{i,A}$; this ceases to be a serious difficulty when, for example, $\mathscr{F}$ is expanded as a series in the strains E_{AB} and truncated after a certain term [cf. Eq. (3.52)].

Toupin [*1*], pp. 884—885, instead of using the variables $\overline{\mathsf{M}}_A$, showed that F would satisfy the rotation condition if it was a function of the strains E_{AB} (or of the C_{AB}'s) and of the three quantities

$$\varPi_A \equiv \mathsf{M}_i\, x_{i,A}. \tag{6.33}$$

Since

$$\overline{\mathsf{M}}_A = \mathsf{M}_i R_{Ai} = \mathsf{M}_i x_{i,B} (C^{-\frac{1}{2}})_{AB} = \varPi_B (C^{-\frac{1}{2}})_{AB}, \tag{6.34}$$

and since the components of $C^{-\frac{1}{2}}$ are functions of the components of C or of E, our function $\mathscr{F}(\overline{\mathsf{M}}_A, C_{PQ})$ is equivalent to a function of the $\varPi_A$'s and C_{PQ}'s. For practical mathematical manipulation, Toupin's $\varPi_A$'s are more convenient variables than our $\overline{\mathsf{M}}_A$'s; but they lack the direct physical significance of the latter. A microscopic theory will lead naturally to expressions involving the $\overline{\mathsf{M}}_A$'s rather than the $\varPi_A$'s.

Chapter III

The Energy Method

7. Formal Theory

7.1. The minimization principle. We now make a new start. We discard the concept of stress, and with it a number of postulates — some of which were stated explicitly, others of which may have been tacitly implied — about physical or metaphysical situations at experimentally inaccessible points of a body. Instead, we endeavor to derive the equations of equilibrium by the thermodynamic minimization principle that was formulated in § 4.3. The principle was applied in § 4.4 to a rigid magnetic body and in § 4.5 to an elastic nonmagnetic body; we must apply it now to a body that is both magnetizable and deformable. Our first problem is to find the proper thermodynamic potential, the G of Eqs. (4.9) and (4.16), for minimization under given applied fields and forces.

We suppose for definiteness that the current I in the magnetizing coil, the body force $\boldsymbol{f}\,dm$ acting on a mass element dm, and the surface force $\boldsymbol{T}'\,dS_0 = T\,dS$ acting on an element of undistorted area dS_0 (corresponding to a constant group of surface particles) are held constant. Then the work in a small change, by all the external forces acting on the whole body, is

$$\delta W = (I/c)\,\delta\Phi_{\rm m} + \int \varrho_0\, f_i\, \delta x_i\, d\tau_0 + \int T_i'\, \delta x_i\, dS_0. \tag{7.1}$$

Here $\Phi_{\rm m}$ is the flux thru the (fixed) magnetizing coil due to the magnetization of the body, $d\tau_0 (\equiv dX_1\, dX_2\, dX_3)$ and dS_0 are elements of undistorted volume and surface, and δx_i is a variation of the i coordinate of the particle whose undeformed coordinates are (X_1, X_2, X_3). Explicitly, if $x_i(X_A)$ are the deformed coordinates of the particle whose undeformed coordinates are X_A, and if $x_i'(X_A)$ are the new deformed coordinates after a small change, then

$$\delta x_i(X_A) \equiv x_i'(X_A) - x_i(X_A). \tag{7.2}$$

It follows that

$$\frac{\partial}{\partial X_A}(\delta x_i) = \frac{\partial x_i'}{\partial X_A} - \frac{\partial x_i}{\partial X_A} = \delta\left(\frac{\partial x_i}{\partial X_A}\right), \tag{7.3}$$

or

$$(\delta x_i)_{,A} = \delta(x_{i,A}) \equiv \delta x_{i,A}: \tag{7.4}$$

the operations δ and $_{,A}$ commute, and the symbol $\delta x_{i,A}$ involves no ambiguity. We may also change the independent variables of the functions δx_i from the X_A's to the x_i's; then

$$\delta x_{i,j} \equiv \frac{\partial}{\partial x_j}(\delta x_i) = \delta x_{i,A}\, X_{A,j}. \tag{7.5}$$

A similar relation connects $\delta \mathsf{M}_{i,j}$ and $\delta \mathsf{M}_{i,A}$.

In the inequality (4.3), we may now replace $P_\alpha\,\delta q_\alpha$ by δW of Eq. (7.1). Thus in natural changes

$$\delta U < (I/c)\,\delta\Phi_{\mathrm{m}} + \int \varrho_0 f_i\,\delta x_i\,d\tau_0 + \int T_i'\,\delta x_i\,dS_0 + T\,\delta\eta. \tag{7.6}$$

At constant T, I, f_i, and T_i', this becomes

$$\delta G < 0, \tag{7.7}$$

where

$$\left.\begin{aligned} G &= U - T\eta - (I/c)\,\Phi_{\mathrm{m}} - \int \varrho_0 f_i\,x_i\,d\tau_0 - \int T_i'\,x_i\,dS_0 \\ &= F - \int \boldsymbol{H}_0\cdot\mathbf{M}\,dm - \int \varrho_0 f_i\,x_i\,d\tau_0 - \int T_i'\,x_i\,dS_0; \end{aligned}\right\} \tag{7.8}$$

the transformation from $(I/c)\,\Phi_{\mathrm{m}}$ to $\int \boldsymbol{H}_0\cdot\boldsymbol{M}\,d\tau = \int \boldsymbol{H}_0\cdot\mathbf{M}\,dm$ is made as in Eq. (2.47), and $F = U - T\eta$ is the Helmholtz function, or internal free energy.

It follows that for stable equilibrium at given T, $\boldsymbol{H}_0$, $\boldsymbol{f}$, and $\boldsymbol{T}'$, G as defined by Eq. (7.8) must be a minimum with respect to internal parameters, such as the M_i and x_i of an arbitrary particle.

To apply this principle, we must have an expression for F in terms of these internal parameters. If we take as internal parameters only the M_i's and x_i's, we are thereby supposing that our expression for F holds when the body is in internal equilibrium but not necessarily in equilibrium with the external generalized forces H_{0i}, f_i, and T_i'.

On the basis of our study of rigid magnetic bodies, we must include in F a term that takes account of long-range magnetic dipole-dipole interactions. For this term we may use either of the quantities W_{m} and W_{m}' of Eqs. (2.50) and (2.52). For definiteness we choose W_{m}. In terms of the undeformed coordinates X_A, W_{m} is given by

$$W_{\mathrm{m}} = -\tfrac{1}{2}\int \varrho_0\,\boldsymbol{H}_1\cdot\mathbf{M}\,d\tau_0, \tag{7.9}$$

with $\boldsymbol{H}_1$ evaluated at the deformed position x_i of the particle whose undeformed coordinates were X_A. To this term in F, we add the integral over the body of a "local" free energy per unit mass:

$$F = W_{\mathrm{m}} + \int \mathsf{F}\,dm = W_{\mathrm{m}} + \int \varrho_0\,\mathsf{F}\,d\tau_0, \tag{7.10}$$

where F is a function of local variables. These variables must include the components of moment per unit mass, M_i, and the deformation gradients $x_{i,A}$; in addition, to take account of exchange forces, we must include the gradients of M_i:

$$\mathsf{F} = \mathsf{F}(\mathsf{M}_i, x_{i,A}, \mathsf{M}_{i,A}). \tag{7.11}$$

We shall thruout this chapter take account of exchange forces. In Chapter II, for simplicity, they were omitted. They could of course have been included there also; for a treatment by the stress method that takes them into account, see TIERSTEN [*1*].

Inclusion of exchange forces means that we are now dealing with a ferromagnetic material (extension of the treatment to include materials with more than one magnetic sublattice, *e.g.*, antiferromagnetic materials, is not within the scope of this monograph). It therefore possesses a spontaneous magnetic moment M_s per unit mass. The magnitude of M_s depends on the temperature and on a quantity known as the exchange integral, which determines the energy of exchange interaction between nearest-neighbor spins in the lattice. The exchange integral depends on the distance between the atoms whose spins it couples; it therefore depends on the strain. Strictly, therefore, we should suppose that the spontaneous magnetic moment per unit mass, M_s, is a function of the strain as well as the temperature; it must then vary from point to point in a body that is nonuniformly strained. This effect will be small at low temperatures, since at $T=0$ the spins are perfectly alined for any positive value of the exchange integral. We shall therefore simplify the problem by supposing that M_s is a function of temperature only, and independent of the strain. Since perfect alinement of spins produces a definite magnetic moment per unit mass, not per unit volume, it is $|\mathbf{M}|$ and not $|\boldsymbol{M}|$ that must be supposed constant; obviously $|\boldsymbol{M}| = \varrho|\mathbf{M}| = \varrho \mathsf{M}_s = \varrho_0 \mathsf{M}_s/J$, where J is the Jacobian $\partial(x_1, x_2, x_3)/\partial(X_1, X_2, X_3)$ [Eq. (3.22)].

We must eventually impose the condition that F is not altered by a rigid rotation of the mass element together with the magnetic moments of all its particles; this condition, however, as was shown by TOUPIN [*1*], is more easily imposed *after* the variational procedure has been carried out.

To find the equilibrium conditions, we compute the variation of G, as given by Eq. (7.8) with F given by Eqs. (7.10), (7.9), and (7.11), for virtual variations $\delta \mathsf{M}_i$ and δx_i that are arbitrary functions, except for the constraint $\mathsf{M}_i \mathsf{M}_i = \mathsf{M}_s^2$, of the undeformed coordinates X_A. To take account of the constraint, we introduce Lagrangian multipliers $\lambda'(X_A)$ in V_0 and $\mu'(X_A)$ on S_0 and require that

$$\delta G - \int \lambda' \mathsf{M}_i \, \delta \mathsf{M}_i \, d\tau_0 - \int \mu' \mathsf{M}_i \, \delta \mathsf{M}_i \, d S_0 = 0 \tag{7.12}$$

for *arbitrary* functions $\delta \mathsf{M}_i$ and δx_i. If the integrals can be so transformed that they contain only the variations $\delta \mathsf{M}_i$ and δx_i, and not their derivatives $\delta \mathsf{M}_{i,A}$ and $\delta x_{i,A}$, then the coefficients of $\delta \mathsf{M}_i$ and δx_i in V_0 and on S_0 may be equated to zero. It is this transformation that is the only difficult step in the variational procedure.

On substituting Eqs. (7.9)—(7.11) in Eq. (7.8), we get for the quantity to be minimized

$$\left.\begin{aligned} G = \int \varrho_0 \mathsf{F}(\mathsf{M}_i, x_{i,A}, \mathsf{M}_{i,A}) \, d\tau_0 - \tfrac{1}{2} \int \varrho_0 \boldsymbol{H}_1 \cdot \mathbf{M} \, d\tau_0 - \\ - \int \varrho_0 \boldsymbol{H}_0 \cdot \mathbf{M} \, d\tau_0 - \int \varrho_0 f_i \, x_i \, d\tau_0 - \int T_i' \, x_i \, d S_0 . \end{aligned}\right\} \tag{7.13}$$

7.2. The magnetic equilibrium conditions. Since δG is linear in the magnetic variations $\delta \mathsf{M}_i$ and the mechanical variations δx_i, we may consider separately the effects of varying M_i at constant x_i and of varying x_i at constant M_i. In this section we consider the former. The conditions obtained by equating the coefficients of $\delta \mathsf{M}_i$ to zero may be called the *magnetic* equilibrium conditions.

In variations of $\delta \mathsf{M}_i$ at constant δx_i, the situation is the same as in a rigid body in which the unvaried x_i's are the permanent coordinates of the particles; we may therefore go over to the x_i's as variables of integration and ignore the undeformed coordinates X_A. The quantity to be varied is then

$$G_1 = \int \varrho \mathsf{F}\, d\tau - \tfrac{1}{2} \int \boldsymbol{H}_1 \cdot \boldsymbol{M}\, d\tau - \int \boldsymbol{H}_0 \cdot \boldsymbol{M}\, d\tau. \tag{7.14}$$

The variation of the first term in G_1 is

$$\delta \int \varrho \mathsf{F}\, d\tau = \int \varrho \left[\frac{\partial \mathsf{F}}{\partial \mathsf{M}_i} \delta \mathsf{M}_i + \frac{\partial \mathsf{F}}{\partial \mathsf{M}_{i,A}} \delta \mathsf{M}_{i,A} \right] d\tau. \tag{7.15}$$

To eliminate $\delta \mathsf{M}_{i,A}$, we transform as follows:

$$\left.\begin{aligned} \int \varrho \frac{\partial \mathsf{F}}{\partial \mathsf{M}_{i,A}} \delta \mathsf{M}_{i,A}\, d\tau &= \int \varrho \frac{\partial \mathsf{F}}{\partial \mathsf{M}_{i,A}} x_{j,A}\, \delta \mathsf{M}_{i,j}\, d\tau \\ &= \int \left\{ \left[\varrho \frac{\partial \mathsf{F}}{\partial \mathsf{M}_{i,A}} x_{j,A}\, \delta \mathsf{M}_i \right]_{,j} - \left[\varrho \frac{\partial \mathsf{F}}{\partial \mathsf{M}_{i,A}} x_{j,A} \right]_{,j} \delta \mathsf{M}_i \right\} d\tau \\ &= \int \varrho \frac{\partial \mathsf{F}}{\partial \mathsf{M}_{i,A}} x_{j,A}\, \delta \mathsf{M}_i\, n_j\, dS - \int \left[\varrho \frac{\partial \mathsf{F}}{\partial \mathsf{M}_{i,A}} x_{j,A} \right]_{,j} \delta \mathsf{M}_i\, d\tau. \end{aligned}\right\} \tag{7.16}$$

Thus

$$\left.\begin{aligned} \delta \int \varrho \mathsf{F}\, d\tau = \int \left\{ \varrho \frac{\partial \mathsf{F}}{\partial \mathsf{M}_i} - \left[\varrho \frac{\partial \mathsf{F}}{\partial \mathsf{M}_{i,A}} x_{j,A} \right]_{,j} \right\} \delta \mathsf{M}_i\, d\tau + \\ + \int \varrho \frac{\partial \mathsf{F}}{\partial \mathsf{M}_{i,A}} x_{j,A}\, \delta \mathsf{M}_i\, n_j\, dS. \end{aligned}\right\} \tag{7.17}$$

The variation of the second term in G_1 can be transformed, just as in Eq. (4.23), to

$$\left.\begin{aligned} \delta\{-\tfrac{1}{2} \int \boldsymbol{H}_1 \cdot \boldsymbol{M}\, d\tau\} &= -\int \boldsymbol{H}_1 \cdot \delta \boldsymbol{M}\, d\tau \\ &= -\int \varrho H_{1i}\, \delta \mathsf{M}_i\, d\tau. \end{aligned}\right\} \tag{7.18}$$

The variation of the third term is

$$\left.\begin{aligned} \delta\{-\int \boldsymbol{H}_0 \cdot \boldsymbol{M}\, d\tau\} &= -\int \boldsymbol{H}_0 \cdot \delta \boldsymbol{M}\, d\tau \\ &= -\int \varrho H_{0i}\, \delta \mathsf{M}_i\, d\tau. \end{aligned}\right\} \tag{7.19}$$

Thus, altogether,

$$\left.\begin{aligned} \delta G_1 = \int \left\{ \varrho \frac{\partial \mathsf{F}}{\partial \mathsf{M}_i} - \left[\varrho \frac{\partial \mathsf{F}}{\partial \mathsf{M}_{i,A}} x_{j,A} \right]_{,j} - \varrho H_i \right\} \delta \mathsf{M}_i\, d\tau + \\ + \int \varrho \frac{\partial \mathsf{F}}{\partial \mathsf{M}_{i,A}} x_{j,A}\, \delta \mathsf{M}_i\, n_j\, dS, \end{aligned}\right\} \tag{7.20}$$

where, as usual,

$$\boldsymbol{H} \equiv \boldsymbol{H}_0 + \boldsymbol{H}_1 . \tag{7.21}$$

The magnetic equilibrium equations are therefore (with $\lambda = \lambda' \, dV_0/dV = J^{-1}\lambda'$, $\mu = \mu' \, dS_0/dS$)

$$\varrho \frac{\partial \mathsf{F}}{\partial \mathsf{M}_i} - \left[\varrho \frac{\partial \mathsf{F}}{\partial \mathsf{M}_{i,A}} x_{j,A}\right]_{,j} - \varrho H_i - \lambda \mathsf{M}_i = 0 \quad \text{in} \quad V, \tag{7.22}$$

$$\varrho \frac{\partial \mathsf{F}}{\partial \mathsf{M}_{i,A}} x_{j,A}\, n_j - \mu \mathsf{M}_i = 0 \quad \text{on} \quad S. \tag{7.23}$$

The first three terms of Eq. (7.22), divided by $-\varrho$, may be regarded as the i component of an "effective field" $\boldsymbol{H}_{\text{eff}}$, which for equilibrium must be in (or opposite to) the direction of $\mathbf{M}$. We can eliminate λ and μ in Eqs. (7.22) and (7.23) by multiplying by M_k and taking the antisymmetric part:

$$\mathsf{M}_{[k}\left\{\varrho \frac{\partial \mathsf{F}}{\partial \mathsf{M}_{i]}} - \left[\varrho \frac{\partial \mathsf{F}}{\partial \mathsf{M}_{i],A}} x_{j,A}\right]_{,j} - \varrho H_{i]}\right\} = 0 \quad \text{in} \quad V, \tag{7.24}$$

$$\varrho \mathsf{M}_{[k} \frac{\partial \mathsf{F}}{\partial \mathsf{M}_{i],A}} x_{j,A}\, n_j = 0 \quad \text{on} \quad S. \tag{7.25}$$

These may be regarded as torque equations: they assert that the volume and surface torques per unit (deformed) volume or surface must vanish in equilibrium. The torque appears in Eqs. (7.24), (7.25) as an antisymmetric tensor rather than as an axial vector; in vector symbols, the torque per unit volume due to $\boldsymbol{H}$ is $\boldsymbol{M} \times \boldsymbol{H} = \varrho \mathbf{M} \times \boldsymbol{H}$, and the total torque per unit volume (which must vanish) is $\varrho \mathbf{M} \times \boldsymbol{H}_{\text{eff}}$.

To express Eq. (7.24) in terms of the undeformed coordinates X_A, we note that in the second term, $\varrho x_{j,A} = \varrho_0 J^{-1} x_{j,A}$ and that [1]

$$(J^{-1} x_{j,A})_{,j} = 0; \tag{7.26}$$

therefore the factors $\varrho x_{j,A}$ can be removed from the brackets. Thus

$$\left.\begin{aligned} -\mathsf{M}_{[k}\left[\varrho \frac{\partial \mathsf{F}}{\partial \mathsf{M}_{i],A}} x_{j,A}\right]_{,j} &= -\mathsf{M}_{[k}\, \varrho x_{j,A} \left[\frac{\partial \mathsf{F}}{\partial \mathsf{M}_{i],A}}\right]_{,j} \\ &= -\mathsf{M}_{[k}\, \varrho \left[\frac{\partial \mathsf{F}}{\partial \mathsf{M}_{i],A}}\right]_{,A} \end{aligned}\right\} \tag{7.27}$$

---

[1] Proof: $(x_{j,A} J^{-1})_{,j} = (x_{j,A})_{,j} J^{-1} + x_{j,A} (J^{-1})_{,j}$. From $J^{-1} = \det X_{A,i}$ and the properties of determinants, $(J^{-1})_{,j} = K_{Bk} (X_{B,k})_{,j} = K_{Bk} X_{B,kj}$, where K_{Bk} is the cofactor of $X_{B,k}$ in J^{-1}. By regarding the three equations $X_{A,i}\, x_{i,B} = \delta_{AB}$ for fixed B and for $A = 1, 2, 3$ as simultaneous equations in the three unknowns $x_{1,B}$, $x_{2,B}$, $x_{3,B}$ and solving, we find $x_{k,B} = J K_{Bk}$; hence $K_{Bk} = J^{-1} x_{k,B}$. Thus $(J^{-1})_{,j} = J^{-1} x_{k\,B} X_{B,kj}$, and $(x_{j,A} J^{-1})_{,j} = J^{-1}\{(x_{j,A})_{,j} + x_{j,A}\, x_{k,B} X_{B,kj}\}$. To evaluate $(x_{j,A})_{,j}$, we apply first $\partial/\partial x_j$ and then $x_{k,A}$ to the identity $x_{j,B} X_{B,k} = \delta_{jk}$; we find $(x_{j,A})_{,j} = -x_{j,B}\, x_{k,A} X_{B,kj} = -x_{k,B}\, x_{j,A} X_{B,jk} = -x_{k,B}\, x_{j,A} X_{B,kj}$, so that the expression in braces vanishes and $(x_{j,A} J^{-1})_{,j} = 0$.

and Eq. (7.24) becomes

$$\mathsf{M}_{[k}\left\{\varrho\frac{\partial \mathsf{F}}{\partial \mathsf{M}_{i]}}-\varrho\left[\frac{\partial \mathsf{F}}{\partial \mathsf{M}_{i],A}}\right]_{,A}-\varrho H_{i]}\right\}=0. \tag{7.28}$$

(If multiplied thru by a factor ϱ_0/ϱ, this becomes Eq. (13) of BROWN [13].) To express Eq. (7.25) in terms of the X_A's, we note that by Eq. (3.28), $n_j\,dS=JX_{C,j}N_C\,dS_0$; therefore $x_{j,A}\,n_j$ may be replaced by a scalar factor times $x_{j,A}X_{C,j}N_C=\delta_{AC}N_C=N_A$, and Eq. (7.25) becomes

$$\mathsf{M}_{[k}\frac{\partial \mathsf{F}}{\partial \mathsf{M}_{i],A}}N_A=0. \tag{7.29}$$

(This is Eq. (14) of BROWN [13].)

Eqs. (2.61) and (2.62) are special cases of Eqs. (7.24) and (7.25), corresponding to the special form of F used in deriving them.

7.3. The mechanical equilibrium conditions. We now vary G, Eq. (7.13), with respect to the three arbitrary functions $\delta x_i(X_A)$, at constant M_i. The resulting formula will involve spatial derivatives $\delta x_{i,A}$ and also a variation of $\boldsymbol{H}_1$, caused by the change of position of the magnetic-moment elements $\mathbf{M}\,dm$. If we can transform so as to eliminate these quantities and get an expression involving only δx_i, we may then equate the coefficient of each $\delta x_i (i=1, 2, 3)$ to zero. The resulting three equations express the mechanical equilibrium conditions.

The variation of the first term in G is

$$\left.\begin{aligned}\delta\int\varrho_0\mathsf{F}\,d\tau_0&=\int\varrho_0\frac{\partial \mathsf{F}}{\partial x_{i,A}}\,\delta x_{i,A}\,d\tau_0=\int\varrho\frac{\partial \mathsf{F}}{\partial x_{i,A}}\,x_{j,A}\,\delta x_{i,j}\,d\tau\\&=\int\left\{\left[\varrho\frac{\partial \mathsf{F}}{\partial x_{i,A}}\,x_{j,A}\,\delta x_i\right]_{,j}-\left[\varrho\frac{\partial \mathsf{F}}{\partial x_{i,A}}\,x_{j,A}\right]_{,j}\delta x_i\right\}d\tau\\&=\int\varrho\frac{\partial \mathsf{F}}{\partial x_{i,A}}\,x_{j,A}\,\delta x_i\,n_j\,dS-\int\left[\varrho\frac{\partial \mathsf{F}}{\partial x_{i,A}}\,x_{j,A}\right]_{,j}\delta x_i\,d\tau.\end{aligned}\right\} \tag{7.30}$$

The variation of the second term, W_{m}, can be obtained (when both M_i and x_i are varied) by replacing $d\mathsf{M}_i/dt$ and v_i in Eq. (6.10) by $\delta\mathsf{M}_i$ and δx_i respectively. The part involving $\delta\mathsf{M}_i$ is easily seen to be the right member of Eq. (7.18). For present purposes we set $\delta\mathbf{M}=0$ and get

$$\delta W_{\mathrm{m}}=-\int\varrho\mathsf{M}_j H_{1i,j}\,\delta x_i\,d\tau-\tfrac{1}{2}\gamma\int M_{\mathrm{n}}^2\,n_i\,\delta x_i\,dS. \tag{7.31}$$

The variation of the third term is

$$\left.\begin{aligned}\delta\{-\textstyle\int\varrho_0\mathsf{M}_j H_{0j}\,d\tau_0\}&=-\textstyle\int\varrho_0\mathsf{M}_j\,\delta H_{0j}\,d\tau_0\\=-\textstyle\int\varrho_0\mathsf{M}_j H_{0j,i}\,\delta x_i\,d\tau_0&=-\textstyle\int\varrho\,\mathsf{M}_j H_{0i,j}\,\delta x_i\,d\tau,\end{aligned}\right\} \tag{7.32}$$

since the variation of H_{0j} is due entirely to the change of position of the mass element $\varrho_0\,d\tau_0$ from x_i to $x_i+\delta x_i$, and since $\nabla\times\boldsymbol{H}_0=0$.

Finally, the variation of the last two terms in G is

$$\left.\begin{aligned}&\delta\{-\textstyle\int\varrho_0 f_i\, x_i\, d\tau_0-\int T_i'\, x_i\, d S_0\}\\ &\quad=-\textstyle\int\varrho_0 f_i\,\delta x_i\, d\tau_0-\int T_i'\,\delta x_i\, d S_0\\ &\quad=-\textstyle\int\varrho f_i\,\delta x_i\, d\tau-\int T_i\,\delta x_i\, d S.\end{aligned}\right\}\tag{7.33}$$

Obviously it would be the same if we had made some other assumption about the behavior of $\boldsymbol{f}$ and $\boldsymbol{T}$ during the variation (cf. § 4.5).

On combining all the terms, we get

$$\left.\begin{aligned}\delta G=&\int\Big\{-\Big[\varrho\frac{\partial \mathrm{F}}{\partial x_{i,A}}x_{j,A}\Big]_{,j}-\varrho\,\mathrm{M}_j H_{1i,j}-\varrho\,\mathrm{M}_j H_{0i,j}-\varrho f_i\Big\}\delta x_i\, d\tau+\\ &+\int\Big\{\varrho\frac{\partial \mathrm{F}}{\partial x_{i,A}}\,x_{j,A}\,n_j-\frac{1}{2}\gamma M_{\mathrm{n}}^2 n_i-T_i\Big\}\,\delta x_i\, d S.\end{aligned}\right\}\tag{7.34}$$

On setting the coefficients of δx_i equal to zero and replacing $H_{0i}+H_{1i}$ by H_i, we get the mechanical equilibrium equations

$$\Big[\varrho\frac{\partial \mathrm{F}}{\partial x_{i,A}}\,x_{j,A}\Big]_{,j}+M_j H_{i,j}+\varrho f_i=0 \quad\text{in}\quad V,\tag{7.35}$$

$$\varrho\frac{\partial \mathrm{F}}{\partial x_{i,A}}\,x_{j,A}\,n_j-\frac{1}{2}\gamma\, M_{\mathrm{n}}^2\, n_i-T_i=0 \quad\text{on}\quad S.\tag{7.36}$$

(These are Eqs. (15) and (16) of BROWN [*13*].)

7.4. Disembodied fields. In the calculation just presented, the "magnetic self-energy" W_{m} can be written

$$W_{\mathrm{m}}=+\tfrac{1}{2}\textstyle\int \mathrm{M}_i\,\varphi_{1,i}\, dm,\tag{7.37}$$

where φ_1 is the scalar potential such that $\boldsymbol{H}_1=-\nabla\varphi_1$ (*i.e.* $H_{1i}=-\varphi_{1,i}$). In the variation

$$\delta W_{\mathrm{m}}=\tfrac{1}{2}\textstyle\int(\mathrm{M}_i\,\delta\varphi_{1,i}+\varphi_{1,i}\,\delta\mathrm{M}_i)\, dm,\tag{7.38}$$

the variations $\delta\varphi_1$ are determined by those of M_i and x_i; they must therefore be eliminated by use of the relation of φ_1 to its sources and by judicious volume-to-surface and surface-to-volume integral transformations. This calculation (given in full in Appendix A) is long and laborious; it was circumvented in § 6.2 by various artifices. TOUPIN [*1*] (and, following him, TIERSTEN [*2*]) circumvented it by another method: introduction of a different potential φ, one that can be varied independently of M_i and x_i. The theoretical basis of this method lies in a variational theorem of magnetostatics, which can be stated as follows (BROWN [*10*], pp. 57—58, Theorem 4 and "companion theorem"; [*8*]).

Let the magnetization of a finite body be a given function $\boldsymbol{M}(x_i)$. Let $\boldsymbol{H}_1$ be the magnetizing force calculated from pole densities $-\nabla\cdot\boldsymbol{M}$ and $\boldsymbol{n}\cdot\boldsymbol{M}$, and let $\boldsymbol{H}'$ be an arbitrary irrotational vector field, regular at infinity. Let

$$W_{MH} \equiv -\int \boldsymbol{M} \cdot \boldsymbol{H}' \, d\tau - \frac{1}{2\gamma} \int_{\text{space}} \boldsymbol{H}'^2 \, d\tau. \tag{7.39}$$

Then in variations of $\boldsymbol{H}'$ under the conditions stated (irrotationality and regularity at infinity), W_{MH} attains its maximum value when $\boldsymbol{H}' = \boldsymbol{H}_1$, and this maximum value is W_{m}.

Obviously this theorem can be restated in terms of potentials φ' and φ_1 such that $\boldsymbol{H}' = -\nabla\varphi'$ and $\boldsymbol{H}_1 = -\nabla\varphi_1$; the irrotationality requirement is then automatically fulfilled.

TOUPIN's procedure is as follows: he replaces W_{m} by W_{MH} in G and requires that the resulting quantity shall have zero first variation when M_1, x_i, and φ' are all varied independently. Clearly if the variation with respect to φ' is performed first, the equations that result are the equations that relate φ_1 to its sources, *i.e.* POISSON's equation and the related boundary conditions; W_{MH} becomes W_{m}; and the rest of the calculation (the variation with respect to M_i and x_i) is the one carried out in § 7.2. The trick that simplifies the calculation is to do the variations not in this order, but simultaneously. The result is the equilibrium equations obtained in §§ 7.2—7.3, together with POISSON's equation (and boundary conditions). The calculation is simpler because the variations of φ' are now independent, instead of being determined by those of M_i and x_i.

We can, if we will, interpret $\boldsymbol{H}' = -\nabla\varphi'$ as a "disembodied" irrotational field that has a self-energy, given by the second term in Eq. (7.39), and an interaction energy with the magnetization, given by the first term. On this interpretation, we must attribute to the disembodied field a high degree of thermodynamic nonconformity, in that it seeks to *maximize*, rather than minimize, its contribution to the thermodynamic potential G. We shall therefore not attempt to interpret $G + W_{MH} - W_{\mathrm{m}}$ thermodynamically.

Because this quantity is, in equilibrium, a minimum with respect to M_i and x_i and a maximum with respect to φ', it can be used only to derive equilibrium conditions and not to test the stability of the equilibrium.[1]

If stability is of interest, the following alternative magnetostatic theorem may be useful. Let $\boldsymbol{B}_1$ be the induction calculated from Amperian current densities $c\nabla \times \boldsymbol{M}$ und $-c\boldsymbol{n} \times \boldsymbol{M}$, and let $\boldsymbol{B}'$ be an arbitrary

[1] TOUPIN [1] and TIERSTEN [2] were interested only in equilibrium conditions (or equations of motion), not in stability tests; they therefore paid no attention to whether their stationarity conditions represented a minimum, a maximum, or something else. In a related tho not identical problem, AHARONI [2] has shown that maximization with respect to φ' followed by minimization with respect to $\boldsymbol{M}$ cannot always be replaced by minimization with respect to $\boldsymbol{M}$ followed by maximization with respect to φ'; whether a similar phenomenon occurs in the present problem, has not been investigated.

solenoidal vector field, regular at infinity. Let

$$W_{MB} \equiv -\int \boldsymbol{M} \cdot \boldsymbol{B}' \, d\tau + \frac{1}{2\gamma} \int_{\text{space}} \boldsymbol{B}'^2 \, d\tau. \tag{7.40}$$

Then in variations of $\boldsymbol{B}'$ under the conditions stated, W_{MB} attains its *minimum* value when $\boldsymbol{B}' = \boldsymbol{B}_1$, and this minimum value is W'_{m}. To apply this theorem, we must replace W_{m} in G by its equivalent $W'_{\mathrm{m}} + \frac{1}{2}\gamma \int \boldsymbol{M}^2 \, d\tau = W'_{\mathrm{m}} + \frac{1}{2}\gamma M_{\mathrm{s}}^2 V$, and then replace W'_{m} by W_{MB}. We can either vary $\boldsymbol{B}'$ under the solenoidality constraints $\nabla \cdot \boldsymbol{B}' = 0$ and $\boldsymbol{n} \cdot (\boldsymbol{B}^+ - \boldsymbol{B}^-) = 0$, or set $\boldsymbol{B}' = \nabla \times \boldsymbol{A}'$ and vary $\boldsymbol{A}'$. We shall not pursue this possibility further, since use of the actual self-field $\boldsymbol{H}_1$ rather than a "disembodied" field involves no serious difficulties.

7.5. Invariance to rotation. Complete generality is not our goal; if it were, we should not be content with an F that depended only on M_i, $x_{i,A}$, and $\mathsf{M}_{i,A}$ but should consider various other possibilities such as dependence on higher derivatives ($x_{i,AB}$ etc.). We shall nevertheless consider, for a while, the most general possible form of F *under the restriction* that it depends only on the variables M_i, $x_{i,A}$, and $\mathsf{M}_{i,A}$.

The function $\mathsf{F}(\mathsf{M}_i, x_{i,A}, \mathsf{M}_{i,A})$ cannot be arbitrary; it must satisfy the physically obvious requirement that the internal free energy $\mathsf{F}\, dm$ of a mass element dm must not be altered by a rigid rotation of the mass element together with the magnetic moments of all its particles.[1] For an F that depends only on M_i, $x_{i,A}$, and $\mathsf{M}_{i,A}$, it is sufficient to require invariance with respect to a rigid rotation of dm together with the four vectors $\mathbf{M}$ and [2] $\partial \mathbf{M}/\partial X_A$ ($A = 1, 2, 3$); and as regards the rigid rotation of dm itself, it is sufficient to specify a rigid rotation of the three vectors $\partial \boldsymbol{r}/\partial X_A$ ($A = 1, 2, 3$).[3] Thus our requirement may finally be stated thus: F is such a function of the seven vectors $\mathbf{M}$, $\partial \boldsymbol{r}/\partial X_A$ ($A = 1, 2, 3$), and $\partial \mathbf{M}/\partial X_A$ ($A = 1, 2, 3$) that it is not affected by a rigid rotation of this system of seven vectors at the point (X_1, X_2, X_3) under consideration.

---

[1] This is of course true only of the *internal* part F of the thermodynamic potential G. The term $-\int \boldsymbol{M} \cdot \boldsymbol{H}_0 \, d\tau$ obviously does not possess the invariance just described; on the contrary, the dependence of this term on orientation gives rise to a torque $\boldsymbol{M} \times \boldsymbol{H}_0 \, dm$ on the element dm, and hence to a tendency of the element to aline itself with its moment along the field direction. Torques may also arise from the mechanical forces $\boldsymbol{f}$ and $\boldsymbol{T}$.

[2] The quantity $\partial \mathbf{M}/\partial X_1 = \lim_{\Delta X_1 \to 0} \frac{1}{\Delta X_1} [\mathbf{M}(X_1 + \Delta X_1, X_2, X_3) - \mathbf{M}(X_1, X_2, X_3)]$ is, for given (X_1, X_2, X_3), a vector with respect to the x_i axes; for $\mathbf{M}(X_1 + \Delta X_1, X_2, X_3)$ and $\mathbf{M}(X_1, X_2, X_3)$ are such vectors, and therefore their difference is such a vector, and this property is not affected by division by ΔX_1 and the limiting process $\Delta X_1 \to 0$.

[3] The components of the vector $\partial \boldsymbol{r}/\partial X_A$ are $x_{i,A}$ (A const; $i = 1, 2, 3$).

In the analogous electric problem, where the only variables in F were the analogs of M_i and $x_{i,A}$, Toupin [1] (pp. 888—889) applied the invariance-to-rotation condition by use of a theorem of Cauchy [1]. The theorem states that an isotropic function of three or more vectors $\boldsymbol{v}^{(\alpha)}$ ($\alpha = 1, 2, \ldots$) — that is, a function invariant to a rigid rotation of the system of vectors — can be expressed as a function of the various quantities $\boldsymbol{v}^{(\alpha)} \cdot \boldsymbol{v}^{(\beta)}$ (where α and β may be the same or different) and $\boldsymbol{v}^{(\alpha)} \cdot \boldsymbol{v}^{(\beta)} \times \boldsymbol{v}^{(\gamma)}$ (where α, β, and γ are all different). Cauchy actually proved the theorem only for three vectors and then stated the generalization without proof; he considered explicitly only position vectors, but generalization in this respect involves no difficulty. The discussion can be simplified by assuming that three of the vectors — say $\boldsymbol{v}^{(1)}$, $\boldsymbol{v}^{(2)}$, and $\boldsymbol{v}^{(3)}$ — are restricted to be noncoplanar, with $\boldsymbol{v}^{(1)} \cdot \boldsymbol{v}^{(2)} \times \boldsymbol{v}^{(3)}$ positive. Then knowledge of their lengths and of the cosines of the angles between them, or equivalently knowledge of the six quantities $\boldsymbol{v}^{(\alpha)} \cdot \boldsymbol{v}^{(\beta)}$ ($\alpha, \beta = 1, 2, 3$), determines the three vectors except for a rigid rotation of them; these six quantities are therefore sufficient arguments for an isotropic function of $\boldsymbol{v}^{(1)}$, $\boldsymbol{v}^{(2)}$, and $\boldsymbol{v}^{(3)}$. For any additional vectors $\boldsymbol{v}^{(\gamma)}$ ($\gamma > 3$), the three quantities $\boldsymbol{v}^{(\gamma)} \cdot \boldsymbol{v}^{(\alpha)}$ ($\alpha = 1, 2, 3$) determine its components in the base system reciprocal to the system $\boldsymbol{v}^{(1)}$, $\boldsymbol{v}^{(2)}$, $\boldsymbol{v}^{(3)}$ and are therefore sufficient additional arguments. Further quantities $\boldsymbol{v}^{(\gamma)} \cdot \boldsymbol{v}^{(\delta)}$ and $\boldsymbol{v}^{(\gamma)} \cdot \boldsymbol{v}^{(\delta)} \times \boldsymbol{v}^{(\varepsilon)}$ are also possible arguments, tho redundant ones.

In the present application, we choose as $\boldsymbol{v}^{(1)}$, $\boldsymbol{v}^{(2)}$, and $\boldsymbol{v}^{(3)}$ the vectors $\partial \boldsymbol{r}/\partial X_1$, $\partial \boldsymbol{r}/\partial X_2$, $\partial \boldsymbol{r}/\partial X_3$ with components $v_i^{(1)} = x_{i,1}$, $v_i^{(2)} = x_{i,2}$, $v_i^{(3)} = x_{i,3}$; then $\boldsymbol{v}^{(1)} \cdot \boldsymbol{v}^{(2)} \times \boldsymbol{v}^{(3)} = \det x_{i,A} = J = dV/dV_0 > 0$. Sufficient arguments for an isotropic function of these three vectors are $\boldsymbol{v}^{(\alpha)} \cdot \boldsymbol{v}^{(\beta)} = v_i^{(\alpha)} v_i^{(\beta)} = x_{i,A}\, x_{i,B} = C_{AB}$ ($\alpha = A = 1, 2, 3$; $\beta = B = 1, 2, 3$). If we allow the function to depend also on $\boldsymbol{v}^{(4)} = \mathbf{M}$, sufficient additional arguments are $\boldsymbol{v}^{(4)} \cdot \boldsymbol{v}^{(\alpha)} = \mathsf{M}_i\, x_{i,A}$ ($\alpha = A = 1, 2, 3$); these are Toupin's variables Π_A. If there were no further independent variables in F, therefore, we should have

$$\mathsf{F} = \mathsf{F}(\Pi_A, C_{AB}), \tag{7.41}$$

with

$$\Pi_A \equiv \mathsf{M}_i\, x_{i,A}, \qquad C_{AB} = x_{i,A}\, x_{i,B} = C_{BA}. \tag{7.42}$$

As we saw in § 6.3, we may use, instead of Toupin's Π_A's, the physically more meaningful quantities $\overline{\mathsf{M}}_A \equiv \mathsf{M}_i R_{A\,i}$, the components of $\mathbf{M}$ in axes that share the rotation $R_{A\,i}$ of the mass element.

If we now allow F to depend also on the vectors $\boldsymbol{v}^{(5)}$, $\boldsymbol{v}^{(6)}$, $\boldsymbol{v}^{(7)}$ with components $\partial \mathsf{M}_i/\partial X_1$, $\partial \mathsf{M}_i/\partial X_2$, $\partial \mathsf{M}_i/\partial X_3$ respectively, sufficient additional arguments are the nine quantities $\boldsymbol{v}^{(\gamma)} \cdot \boldsymbol{v}^{(\alpha)}$ ($\alpha = 1, 2, 3$; $\gamma = 5, 6, 7$), or $\mathsf{M}_{i,A}\, x_{i,B}$ ($A, B = 1, 2, 3$). Other possibilities, in general redundant but in particular situations more convenient, include the six quantities $\boldsymbol{v}^{(\gamma)} \cdot \boldsymbol{v}^{(\delta)}$ ($\gamma, \delta = 5, 6, 7$), or $\mathsf{M}_{i,A}\, \mathsf{M}_{i,B}$ ($A, B = 1, 2, 3$).

We saw in § 2.7, Eqs. (2.57) and (2.58), that for a rigid magnetizable body, exchange forces are taken sufficiently into account by including in the volume integrand of F_{loc} a term $\frac{1}{2} b_{ij}\, \alpha_{k,i}\, \alpha_{k,j}$. The corresponding expression in our present $\varrho_0 \mathsf{F}$ is $\frac{1}{2} K_{AB}\, \alpha_{k,A}\, \alpha_{k,B}$, where K_{AB} must now be a function of the strains (we shall examine this question in more detail in § 8.2). If linear terms in the strains are sufficient, we may set $K_{AB} = b_{AB} + b_{CDAB} E_{CD}$. Then the exchange contribution to F is

$$\left.\begin{aligned} \varrho_0 \mathsf{F}_{\text{ex}} &= \tfrac{1}{2} \{ b_{AB}\, \alpha_{k,A}\, \alpha_{k,B} + b_{CDAB} E_{CD}\, \alpha_{k,A}\, \alpha_{k,B} \} \\ &= (2 \mathsf{M}_s^2)^{-1} \{ b_{AB}\, \mathsf{M}_{i,A}\, \mathsf{M}_{i,B} + b_{ABCD} E_{AB}\, \mathsf{M}_{i,C}\, \mathsf{M}_{i,D} \}. \end{aligned}\right\} \tag{7.43}$$

We see that this expression is a function of the quantities $E_{AB} [= \frac{1}{2}(\delta_{AB} - C_{AB})]$ and $\mathsf{M}_{i,A}\, \mathsf{M}_{i,B}$, in accordance with the requirements of our general theory.

The remaining part of F is a function only of the E_{AB}'s and the Π_A's, or equivalently of the E_{AB}'s and the $\overline{\mathsf{M}}_A$'s. Thus

$$\mathsf{F} = \mathsf{F}_{\text{ex}} + \mathscr{F}(\overline{\mathsf{M}}_P, E_{AB}). \tag{7.44}$$

(Eqs. (7.43) and (7.44) are Eqs. (18) and (19) of Brown [*13*]).

7.6. Comparison with the stress method. In Eqs. (7.24), (7.25) and (7.35), (7.36) we have a set of partial differential equations and boundary conditions that determine the equilibrium functions $\alpha_i(X_A)$ and $x_i(X_A)$. Neither in these equations nor in the theory that led to them does the concept of "stress" enter.

In our previous derivation of equilibrium equations, in Chap. II, the stress concept played a fundamental role; energy arguments were introduced only after equilibrium conditions had been derived, and then only to derive relations between the stress and magnetizing-force components on the one hand and the deformation-gradient and magnetization components on the other.

We shall now compare the two sets of formulas. For exact equivalence, we must in our present derivation omit the exchange energy and remove the constraint $\mathbf{M}^2 = \text{const}$, since these were absent in our stress-method calculations. The result of this change is that the magnetic equilibrium equations become Eq. (7.22) with $\partial \mathsf{F}/\partial \mathsf{M}_{i,A}$ and λ set equal to zero:

$$\frac{\partial \mathsf{F}}{\partial \mathsf{M}_i} - H_i = 0 \quad \text{in } V; \tag{7.45}$$

the boundary condition (7.23) is no longer required. The mechanical equilibrium equations (7.35) and (7.36) are unchanged.

It is at once evident that Eq. (7.45) is identical with the previous equation (6.24). To compare Eqs. (7.35), (7.36) with previous results, we must, in the previous equilibrium equations (5.31) (with $a_i = 0$) and

(5.32), substitute the value of t_{ij} from Eq. (6.25). The results are identical with Eqs. (7.35) and (7.36).

The relation (6.25),

$$t_{ij} = \varrho (\partial \mathsf{F}/\partial x_{i,A})\, x_{j,A}, \tag{7.46}$$

gives formulas for the various "stresses" τ_{ij}, $\bar{t}_{ij}$, $\bar{t}'_{ij}$, and t'_{ij} by insertion of it in the equations that relate them to t_{ij} (§ 5.6). It is clear, therefore, that the energy method provides no reason for preferring one set of "stresses" to another. In fact, it provides no reason for introducing *any* set of stresses; all formulas for observable quantities — the moments and deformations under given applied field $\boldsymbol{H}_0$ and applied forces $\boldsymbol{f}$ and $\boldsymbol{T}$ — can be derived without ever mentioning "stresses".

If we do decide to introduce the stress concept, and if we choose a particular one of the sets of "stresses" defined in § 5.6, these stresses are still not uniquely defined; for in order to reduce the energy-method equilibrium equations (7.35), (7.36) to the stress-method equilibrium equations (5.31), (5.32), we need not necessarily adopt the relation (7.46): the more general relation

$$t_{ij} = \varrho (\partial \mathsf{F}/\partial x_{i,A})\, x_{j,A} + P_{ij}, \tag{7.47}$$

where P_{ij} is subject only to the conditions $P_{ij,j} = 0$ in V, $P_{ij}\, n_j = 0$ on S, and $P_{[ij]} = 0$ in V will do equally well. (The last condition, $P_{[ij]} = 0$, will be discussed in § 7.7). It will be recalled that in the energy arguments of § 6.3, we were able to derive Eq. (6.25) only if we postulated that the internal forces were actually distributed as $t_i(\boldsymbol{n})\, dS$ across an arbitrary internal dS, and not if we postulated only that their forces and moments were equivalent to those of $t_i(\boldsymbol{n})\, dS$ for a closed surface; in the latter case we could derive only the relations (7.47).

Our conclusion from this discussion is that the stress concept is completely dispensable, and that arguments about the relative merits of various sets of "stresses" are futile: this is a matter of arbitrary definition, to be decided on the basis of convenience or of esthetic preference.

The field vector $\boldsymbol{H}_1$ (or $\boldsymbol{H}$) is, in principle, equally dispensable. It was defined in a formal manner; the only physical property postulated for it is the property that $W_{\mathrm{m}} \equiv -\frac{1}{2}\int \boldsymbol{M} \cdot \boldsymbol{H}_1\, d\tau$ contains all the long-range dipole-dipole interactions. That property is possessed also by W'_{m}, in which $\boldsymbol{B}_1$ replaces $\boldsymbol{H}_1$, and by any linear combination $\alpha W_{\mathrm{m}} + (1-\alpha) W'_{\mathrm{m}}$, in which $\alpha \boldsymbol{H}_1 + (1-\alpha)\boldsymbol{B}_1$ replaces $\boldsymbol{H}_1$. The microscopic calculation that demonstrates this property (Brown [*10*], pp. 101—102) uses only the formal definition of $\boldsymbol{H}_1$.

In practice, the nonuse of $\boldsymbol{H}_1$ (or one of its alternates, such as $\boldsymbol{B}_1$) would require direct use of the integrals (2.35) and would be somewhat inconvenient — more so than the omission of stresses. For that reason we have not attempted to dispense with it. We nevertheless regard it as an auxiliary quantity, convenient but not essential; and "physical" interpretations of it, as a supposed actual "field intensity" of some kind, are as superfluous as are "physical" interpretations of one or another system of "stresses".

7.7. The asymmetry of the stresses. We shall hereafter use the notation t_{ij} for the quantity $\varrho(\partial \mathsf{F}/\partial x_{i,A})\, x_{j,A}$, but we shall regard this as primarily a device for abbreviating the equations; we shall not insist on the physical importance of the t_{ij}'s (or the τ_{ij}'s). Introduction of the abbreviation t_{ij} does not change the magnetic equilibrium equations (7.24), (7.25), but it reduces the mechanical equilibrium equations (7.35), (7.36) to the compact form

$$t_{ij,j} + M_i H_{i,j} + \varrho f_i = 0 \quad \text{in } V, \tag{7.48}$$

$$t_{ij} n_j - \tfrac{1}{2}\gamma M_{\mathrm{n}}^2 n_i - T_i = 0 \quad \text{on } S. \tag{7.49}$$

We have still to show that the energy method leads to an antisymmetric part $t_{[ij]}$ of t_{ij} that reduces to Eq. (5.30), $t_{[ij]} = M_{[i} H_{j]}$, when exchange forces are omitted. For this purpose, we will first derive a formula for $t_{[ij]}$ in the presence of exchange forces and then consider the case when they are absent.

As was shown by Toupin [*1*], pp. 884—885, the asymmetry of t_{ij} is determined by the requirement, discussed in § 7.5, that F must be invariant with respect to a rigid rotation of a certain system of vectors $\boldsymbol{v}^{(\alpha)}$. When the arguments of F are M_i, $x_{i,A}$, and $\mathsf{M}_{i,A}$, there are seven such vectors: $\mathbf{M}$, $\partial \boldsymbol{r}/\partial X_A$ ($A = 1, 2, 3$), and $\partial \mathbf{M}/\partial X_A$ ($A = 1, 2, 3$).

For the present purpose, it is sufficient to require this invariance in an *infinitesimal* rotation. In such a rotation, $\delta \boldsymbol{v}^{(\alpha)} = \boldsymbol{\omega} \times \boldsymbol{v}^{(\alpha)}$, or $\delta v_i^{(\alpha)} = \omega_{ij} v_j^{(\alpha)}$, where ω_{ij} is an antisymmetric tensor, equivalent to the axial vector $\boldsymbol{\omega}$; the rotation is thru a small angle ω about the direction $\boldsymbol{\omega}/\omega$.

The requirement is that

$$\delta \mathsf{F} = \frac{\partial \mathsf{F}}{\partial v_i^{(\alpha)}}\, \delta v_i^{(\alpha)} = \frac{\partial \mathsf{F}}{\partial v_i^{(\alpha)}}\, \omega_{ij}\, v_j^{(\alpha)} \tag{7.50}$$

(the sum is over i, j, and α) shall vanish for arbitrary antisymmetric ω_{ij}; it follows that the antisymmetric (in i and j) part of the coefficient of ω_{ij} in the expression (7.50) must vanish for $i = 1, 2, 3$ and $j > i$. Hence

$$\frac{\partial \mathsf{F}}{\partial v_{[i}^{(\alpha)}}\, v_{j]}^{(\alpha)} = 0. \tag{7.51}$$

On insertion of the specific $v_i^{(\alpha)}$'s — namely, M_i, $x_{i,A}$ ($A=1, 2, 3$), and $\mathsf{M}_{i,A}$ ($A=1, 2, 3$), this becomes

$$\frac{\partial \mathsf{F}}{\partial \mathsf{M}_{[i}} \mathsf{M}_{j]} + \frac{\partial \mathsf{F}}{\partial x_{[i,A}} x_{j],A} + \frac{\partial \mathsf{F}}{\partial \mathsf{M}_{[i,A}} \mathsf{M}_{j],A} = 0 \tag{7.52}$$

(in the second and third terms, there is a summation over $A=1, 2, 3$). It follows that

$$t_{[ij]} = \varrho \frac{\partial \mathsf{F}}{\partial x_{[i,A}} x_{j],A} = -\varrho \left\{ \frac{\partial \mathsf{F}}{\partial \mathsf{M}_{[i}} \mathsf{M}_{j]} + \frac{\partial \mathsf{F}}{\partial \mathsf{M}_{[i,A}} \mathsf{M}_{j],A} \right\}. \tag{7.53}$$

When exchange forces are omitted, the terms containing $\mathsf{M}_{i,A}$ are absent; also the magnetic equilibrium equation (7.22) reduces to $\partial\mathsf{F}/\partial\mathsf{M}_i = H_i$. Eq. (7.53) then becomes

$$t_{[ij]} = -\varrho H_{[i} \mathsf{M}_{j]} = \varrho \mathsf{M}_{[i} H_{j]} = M_{[i} H_{j]}, \tag{7.54}$$

in agreement with Eq. (5.30).

With exchange forces present, the magnetic equilibrium equation (7.24) enables us to replace $(\partial\mathsf{F}/\partial\mathsf{M}_{[i})\mathsf{M}_{j]}$ by $H_{[i}\mathsf{M}_{j]}$ plus additional terms, and there are other additional terms in Eq. (7.53). Therefore $M_{[i}H_{j]}$ is now only one of several terms in $t_{[ij]}$. Physically, this means that exchange forces as well as magnetic forces can produce a couple proportional to the size of the volume element under consideration. We shall not investigate the additional terms in $t_{[ij]}$ by the present method; the asymmetric terms in t_{ij} will appear automatically when t_{ij} is derived from a function F that satisfies the rotation requirement.

Since the rotation requirement determines the asymmetric part of $\varrho(\partial\mathsf{F}/\partial x_{i,A})\,x_{j,A}$ in the theory based on the energy method, and since this is identical with the asymmetric part of t_{ij} as derived by the stress method, evidently the asymmetric part of P_{ij} in Eq. (7.47) must be zero, as was stated in the discussion of that equation.

7.8. Dynamic modifications. The equilibrium equations can be most easily transformed into dynamic equations by the following method.

The magnetic volume equilibrium equation (7.24) is equivalent to

$$\mathbf{M} \times \boldsymbol{H}_{\text{eff}} = 0, \tag{7.55}$$

where

$$(H_{\text{eff}})_i = H_i - \frac{\partial \mathsf{F}}{\partial \mathsf{M}_i} + \varrho^{-1}\left[\varrho \frac{\partial \mathsf{F}}{\partial \mathsf{M}_{i,A}} x_{j,A}\right]_{,j}; \tag{7.56}$$

it asserts that for equilibrium, the torque $\mathbf{M} \times \boldsymbol{H}_{\text{eff}}$ acting on the magnetic moment of unit mass must vanish. The equation of motion of the moment per unit mass, when the only torque is that due to an applied field $\boldsymbol{H}_0$, is

$$\frac{d\mathbf{M}}{dt} = \gamma_0 \mathbf{M} \times \boldsymbol{H}_0, \tag{7.57}$$

where γ_0 is the ratio of magnetic moment to angular momentum for the moment carriers (electrons). To take account of additional torques, we replace $\boldsymbol{H}_0$ by $\boldsymbol{H}_{\text{eff}}$. Thus the magnetic equation of motion is

$$\frac{d\mathbf{M}}{dt} = \gamma_0 \mathbf{M} \times \boldsymbol{H}_{\text{eff}}. \tag{7.58}$$

This procedure is discussed in more detail in BROWN [*11*], pp. 24—26.

The mechanical volume equilibrium equation (7.35) or (7.48) is equivalent to

$$\boldsymbol{f}_{\text{eff}} = 0, \tag{7.59}$$

where

$$(f_{\text{eff}})_i = f_i + \mathsf{M}_j H_{i,j} + \varrho^{-1} t_{ij,j}; \tag{7.60}$$

it asserts that for equilibrium, the force $\boldsymbol{f}_{\text{eff}}$ acting on unit mass must vanish. The equation of motion of unit mass, when the only force is the "body force" $\boldsymbol{f}$, is

$$\varrho \boldsymbol{a} = \varrho \boldsymbol{f}, \tag{7.61}$$

where $\boldsymbol{a}$ is the acceleration. To take account of additional forces, we replace $\boldsymbol{f}$ by $\boldsymbol{f}_{\text{eff}}$. Thus the mechanical equation of motion is

$$\varrho \boldsymbol{a} = \varrho \boldsymbol{f}_{\text{eff}}. \tag{7.62}$$

(This is equivalent to the equation of motion (5.31) derived by the stress method.)

The magnetic and mechanical surface equations (7.25) and (7.36) [or (7.49)] require no modification, since there is no surface concentration of mass or magnetic moment.

The dynamic equations thus obtained will be valid as long as the time rate or frequency of the changes is not great enough to perturb appreciably the isothermal character of the processes.

A more sophisticated method of deriving dynamic equations is to treat G as the potential energy W of a dynamic system, form the kinetic energy $\mathcal{T}$ and Lagrangian $\mathcal{L} = \mathcal{T} - W$, and apply HAMILTON's principle,

$$\delta \int_0^{t_1} \mathcal{L}\, dt = 0. \tag{7.63}$$

For details of the magnetic aspect of this procedure, see BROWN [*11*], Chap. 3. The mechanical aspect is simpler; the kinetic energy is $\frac{1}{2}\int \varrho_0 \dot{x}_i \dot{x}_i \, d\tau_0$, where $\dot{x}_i$ is the derivative of x_i with respect to time at constant (X_1, X_2, X_3), and the variational procedure is that carried out for infinitesimal strains by LOVE [*1*], pp. 166—167. The transformations required in the potential energy variation are those carried out in § 7.3, and the transformations required in the kinetic energy variation are elementary.

The time rate of change of momentum of the whole body is

$$\frac{d}{dt}\int \dot{x}_i \, dm = \int \ddot{x}_i \, dm = \int \varrho a_i \, d\tau = \int (\varrho f_i + M_j H_{i,j} + t_{ij,j}) \, d\tau, \quad (7.64)$$

by Eqs. (7.60) and (7.62). The last term is

$$\int t_{ij,j} \, d\tau = \int t_{ij} n_j \, dS = \int (T_i + \tfrac{1}{2} \gamma M_{\rm n}^2 n_i) \, dS, \quad (7.65)$$

by Eq. (7.49). Hence

$$\frac{d}{dt}\int \dot{x}_i \, dm = \int \varrho f_i \, d\tau + \int T_i \, dS + (F_{\rm mag})_i, \quad (7.66)$$

where

$$(F_{\rm mag})_i = \int M_j H_{i,j} \, d\tau + \tfrac{1}{2} \gamma \int M_{\rm n}^2 n_i \, dS \quad (7.67)$$

is the magnetic force on the whole body, according to Eq. (5.15) applied to the whole body. We can write $\boldsymbol{F}_{\rm mag}$ more simply in its original form (5.3),

$$\boldsymbol{F}_{\rm mag} = \int \boldsymbol{M} \cdot \nabla \boldsymbol{H}_0 \, d\tau. \quad (7.68)$$

The time rate of change of *mechanical* angular momentum is

$$\frac{d}{dt}\int \boldsymbol{r} \times \dot{\boldsymbol{r}} \, dm = \int \boldsymbol{r} \times \ddot{\boldsymbol{r}} \, dm = \int \boldsymbol{r} \times \boldsymbol{a} \, dm. \quad (7.69)$$

We can represent angular momentum as an antisymmetric tensor $\int x_{[i} \dot{x}_{j]} \, dm = \frac{1}{2} \int (x_i \dot{x}_j - x_j \dot{x}_i) \, dm$ rather than a vector $\int \boldsymbol{r} \times \dot{\boldsymbol{r}} \, dm$; then we have

$$\frac{d}{dt}\int x_{[i} \dot{x}_{j]} \, dm = \int x_{[i} a_{j]} \, dm = \int x_{[i} \{\varrho f_{j]} + M_k H_{j],k} + t_{j]k,k}\} \, d\tau, \quad (7.70)$$

by Eqs. (7.60) and (7.62). The last term is

$$\left.\begin{aligned} \int x_{[i} t_{j]k,k} \, d\tau &= \int [(x_{[i} t_{j]k})_{,k} - x_{[i,k} t_{j]k}] \, d\tau \\ &= \int x_{[i} t_{j]k} n_k \, dS - \int \delta_{[ik} t_{j]k} \, d\tau \\ &= \int x_{[i} \{T_{j]} + \tfrac{1}{2} \gamma M_{\rm n}^2 n_{j]}\} \, dS - \int t_{[ji]} \, d\tau, \end{aligned}\right\} \quad (7.71)$$

by Eq. (7.49). Hence

$$\left.\begin{aligned} \frac{d}{dt}\int x_{[i} \dot{x}_{j]} \, dm &= \int x_{[i} \varrho f_{j]} \, d\tau + \int x_{[i} T_{j]} \, dS + \\ &+ \int x_{[i} M_k H_{j],k} \, d\tau + \frac{1}{2} \gamma \int x_{[i} M_{\rm n}^2 n_{j]} \, dS + \int t_{[ij]} \, d\tau. \end{aligned}\right\} \quad (7.72)$$

The third and fourth integrals are the tensor representation of the torque due to $\boldsymbol{F}_{\rm mag}$. They differ from the magnetic torque on the whole body [see remarks after Eq. (5.15)] by not including the couple term $\int M_{[i} H_{j]} \, d\tau$.

The time rate of change of the gyromagnetic angular momentum

$$\boldsymbol{G} = \gamma_0^{-1} \int \mathbf{M} \, dm \quad (7.73)$$

associated with the magnetic moment of the body is, by Eq. (7.58), $\int \mathsf{M} \times \boldsymbol{H}_{\text{eff}}\, dm$. Again using the tensor representation and inserting the the value of $(H_{\text{eff}})_j$ from Eq. (7.56), we have

$$\left.\begin{aligned}\frac{dG_{ij}}{dt} &= \int \mathsf{M}_{[i} (H_{\text{eff}})_{j]}\, dm \\ &= \int \mathsf{M}_{[i} \left\{ H_{j]} - \frac{\partial \mathsf{F}}{\partial \mathsf{M}_{j]}} + \varrho^{-1} \left[\varrho \frac{\partial \mathsf{F}}{\partial \mathsf{M}_{j],A}} x_{k,A} \right]_{,k} \right\} dm \\ &= \int \mathsf{M}_{[i} \left\{ \varrho H_{j]} - \varrho \frac{\partial \mathsf{F}}{\partial \mathsf{M}_{j]}} + \left[\varrho \frac{\partial \mathsf{F}}{\partial \mathsf{M}_{j],A}} x_{k,A} \right]_{,k} \right\} d\tau .\end{aligned}\right\} \tag{7.74}$$

The last term is

$$\left.\begin{aligned}&\int \mathsf{M}_{[i} \left[\varrho \frac{\partial \mathsf{F}}{\partial \mathsf{M}_{j],A}} x_{k,A} \right]_{,k} d\tau \\ &\quad = \int \mathsf{M}_{[i}\, \varrho \frac{\partial \mathsf{F}}{\partial \mathsf{M}_{j],A}} x_{k,A}\, n_k\, dS - \int \mathsf{M}_{[i,k}\, \varrho \frac{\partial \mathsf{F}}{\partial \mathsf{M}_{j],A}} x_{k,A}\, d\tau \\ &\quad = - \int \varrho\, \mathsf{M}_{[i,A} \frac{\partial \mathsf{F}}{\partial \mathsf{M}_{j],A}}\, d\tau ;\end{aligned}\right\} \tag{7.75}$$

the surface integrand vanishes, by Eq. (7.25). Hence

$$\left.\begin{aligned}\frac{dG_{ij}}{dt} &= \int \left\{ M_{[i} H_{j]} - \varrho \mathsf{M}_{[i} \frac{\partial \mathsf{F}}{\partial \mathsf{M}_{j]}} - \varrho \mathsf{M}_{[i,A} \frac{\partial \mathsf{F}}{\partial \mathsf{M}_{j],A}} \right\} d\tau \\ &= \int \{ M_{[i} H_{j]} - t_{[ij]} \}\, d\tau ,\end{aligned}\right\} \tag{7.76}$$

by Eq. (7.53).

On adding Eqs. (7.72) and (7.76), we get

$$\left.\begin{aligned}\frac{d}{dt}\left\{ \int x_{[i} \dot{x}_{j]}\, dm + G_{ij} \right\} &= \int x_{[i}\, \varrho f_{j]}\, d\tau + \int x_{[i} T_{j]}\, dS + \\ &\quad + \int x_{[i} M_k H_{j],k}\, d\tau + \tfrac{1}{2} \gamma \int x_{[i} M_{\text{n}}^2 n_{j]}\, dS + \int M_{[i} H_{j]}\, d\tau \\ &= \int x_{[i}\, \varrho f_{j]}\, d\tau + \int x_{[i} T_{j]}\, dS + (L_{\text{mag}})_{ij} ,\end{aligned}\right\} \tag{7.77}$$

where $\boldsymbol{L}_{\text{mag}}$ is the whole magnetic torque, including the couple term $\int \boldsymbol{M} \times \boldsymbol{H}\, d\tau$. We can write $\boldsymbol{L}_{\text{mag}}$ more simply in its original form (5.5):

$$\boldsymbol{L}_{\text{mag}} = \int \boldsymbol{r} \times (\boldsymbol{M} \cdot \nabla \boldsymbol{H}_0)\, d\tau + \int \boldsymbol{M} \times \boldsymbol{H}_0\, d\tau . \tag{7.78}$$

7.9. Permissible equilibrium specifications of applied forces. In equilibrium, Eqs. (7.66), (7.72), and (7.76) become

$$\int \varrho f_i\, d\tau + \int T_i\, dS + (F_{\text{mag}})_i = 0 , \tag{7.79}$$

$$\left.\begin{aligned}\int x_{[i}\, \varrho f_{j]}\, d\tau + \int x_{[i} T_{j]}\, dS + \int x_{[i} M_k H_{j],k}\, d\tau + \\ + \tfrac{1}{2} \gamma \int x_{[i} M_{\text{n}}^2 n_{j]}\, dS + \int t_{[ij]}\, d\tau = 0 ,\end{aligned}\right\} \tag{7.80}$$

$$\int \{ M_{[i} H_{j]} - t_{[ij]} \}\, d\tau = 0 . \tag{7.81}$$

Addition of Eqs. (7.80) and (7.81) gives

$$\int x_{[i}\,\varrho\, f_{j]}\, d\tau + \int x_{[i}\, T_{j]}\, d S + (L_{\mathrm{mag}})_{ij} = 0. \tag{7.82}$$

For a nonmagnetic body, the specified body and surface forces must produce zero total force and torque if an equilibrium solution is to be possible. For a magnetic body, these conditions are replaced by Eqs. (7.79) and (7.82). Equilibrium solutions are now possible in which the specified body and surface forces produce a non-vanishing resultant force or torque (in the case of uniform $\boldsymbol{H}_0$, only a torque), equilibrated by the magnetic force $\boldsymbol{F}_{\mathrm{mag}}$ or torque $\boldsymbol{L}_{\mathrm{mag}}$. Except in the case of uniform $\boldsymbol{H}_0$, when we can still require zero resultant force, the new relations (7.79) and (7.82) do not enable us to determine in advance what specifications of f_i and T_i are compatible with equilibrium; for $\boldsymbol{F}_{\mathrm{mag}}$ (except when $\boldsymbol{H}_0$ is uniform) and $\boldsymbol{L}_{\mathrm{mag}}$ depend on the unitially unknown magnetization distribution [see Eqs. (7.67) and (7.78)]. The procedure for finding a solution must now be as follows. Assume some $\boldsymbol{H}_0$, f_i, and T_i, and try to find a solution of the equilibrium equations (7.24), (7.25) and (7.35), (7.36). If a solution exists, it will automatically satisfy Eqs. (7.79) and (7.82); if no solution exists, the assumed $\boldsymbol{H}_0$, f_i, and T_i are not compatible with equilibrium.

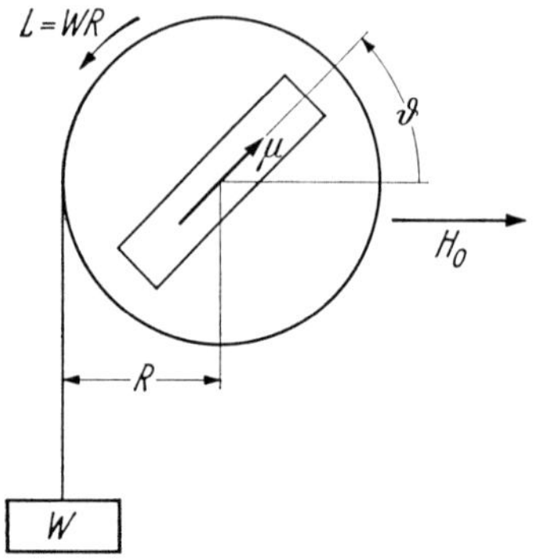

Fig. 13. System with permanent magnetic moment μ, subject to applied magnetic field H_0 and torque L. Not all sets of values of L and H_0 are compatible with the equilibrium condition (7.84)

The situation can be illustrated by the following simple case (Fig. 13). Suppose that a magnet, with permanent magnetic moment μ along its axis, is capable of rotation about an axis perpendicular to a uniform applied field $\boldsymbol{H}_0$. Then if ϑ is the angle from the field direction to the moment direction and if L is a constant applied torque about the ϑ axis, the work done in a small change is $H_0\,\delta(\mu\cos\vartheta) = L\,\delta\,\vartheta$. By the usual thermodynamic reasoning, we find that for stable equilibrium at given T, H_0, and ϑ, we must minimize

$$G = U - T\eta - H_0\,\mu\cos\vartheta - L\vartheta. \tag{7.83}$$

Hence $\partial G/\partial\vartheta = 0$ (for equilibrium) and $\partial^2 G/\partial\vartheta^2 > 0$ (for stability); or

$$\sin\vartheta = L/H_0\,\mu \quad \text{(equilibrium)} \tag{7.84}$$

and

$$H_0\,\mu\cos\vartheta > 0 \quad \text{(stability)}. \tag{7.85}$$

If we have specified an L and H_0 such that $|L| \leqq |H_0\,\mu|$, a solution exists; if $|L| > |H_0\, m|$, no equilibrium is possible, but there will be a constant (in time) angular acceleration. It was necessary to solve the equilibrium equation and obtain a formula for the equilibrium ϑ in order to find the permissible range for L at given $\boldsymbol{H}_0$. The situation in the case of a magnetizable and deformable body is similar: we must solve the equilibrium problem in order to find the permissible sets of values for f_i and T_i. The relations (7.79) and (7.82) are therefore considerably less useful than their analogs for a nonmagnetic body. An exception is Eq. (7.79) in the case of uniform $\boldsymbol{H}_0$.

These results have been deduced by the energy method, without a physical interpretation of t_{ij}. If, however, we adopt our original definition of magnetic forces and hence of stresses τ_{ij} (§§ 5.4, 5.5), then from Eqs. (5.28) we see that $t_{[ij]}=\tau_{[ij]}$. Eq. (7.81) then states that in equilibrium, the volume average of the antisymmetric part of the stress array (not a tensor!), as we have defined it, is equal to the volume average of the tensor that represents $\boldsymbol{M}\times\boldsymbol{H}$. This is not true of the values at a point, because the exchange forces also contribute; but their contributions cancel for the body as a whole, since they are mutual internal forces.

8. Terms in the Free Energy

8.1. Relevance of microscopic concepts. In §7.5 we considered the most general form of the local free energy F per unit mass, under the restrictions that (1) it is to be a function of the quantities M_i, $x_{i,A}$, and $\mathsf{M}_{i,A}$ and that (2) it must be invariant with respect to a rigid rotation of the mass element dm together with the magnetic moments of all its particles. We shall now investigate the more specialized form (7.44) of F (where F_{ex} is given by Eq. (7.43). We shall also look more thoroly into the adequacy of W_{m}, Eq. (7.9), for taking account of long-range dipole-dipole interactions.

The discussion of these matters will not be purely phenomenological but will be based to some extent on microscopic concepts and models. In our primarily phenomenological theory, the function of microscopic models is not to provide final formulas but to guide us in the selection of variables and of mathematical forms. This can be illustrated by the case of exchange forces. A simple exchange-interaction model leads, for a rigid cubic crystal, to a free-energy density term of the form (2.58); this suggests, more generally, Eq. (2.57), where the energy density contains a term $\frac{1}{2}b_{ij}\,\alpha_{k,i}\,\alpha_{k,j}$ quadratic in the *gradients* of the direction cosines α_i. The essential contribution of the microscopic model to our phenomenological theory is the idea of including the gradients $\alpha_{k,i}$ of the direction cosines, in addition to the direction cosines α_i themselves, as arguments in F. In principle, this idea might have occurred to us without any help from a microscopic model or from the quantum-theoretical concept of exchange forces; but in fact, it seems not to have occurred to the illustrious scientists of the century before 1925. Furthermore, there would have been little incentive to complicate the theory with such an idea in the absence of any indications of its physical relevance. But having introduced the idea, we can select proper arrays of b_{ij}'s for various crystals by phenomenological methods, without bothering to work out the corresponding microscopic calculations; the role of more sophisticated models will then be to give us, ultimately, numerical values of the parameters in our phenomenological theory.

The practical sequence of procedures is, then: (1) draw freely on microscopic models, be they ever so crude, as a means of deciding what types of expression, and what independent variables, are likely to be important in the internal free energy (Helmholtz function) F; (2) use phenomenological methods to deduce precise forms for these expressions, and to deduce equilibrium conditions and equations of motion; (3) return ultimately to microscopic theory, but now with more realistic models, to calculate numerical values of the parameters of stage (2). The present monograph is concerned primarily with stage (2); it uses the results of stage (1) and leaves stage (3) for later consideration.

8.2. The exchange energy. The usual derivation of Eq. (2.58) for a *rigid body* (Brown [*10*], pp. 133—135) can be generalized as follows to a deformable body.

We suppose that the spins are localized on atoms at sites of a lattice, and that the exchange-interaction energy is negligible except between nearest neighbors. For nearest-neighbor sites α and β, we take the interaction energy to be

$$W^{(\alpha\beta)} = -2J(r^{(\alpha\beta)})\, \boldsymbol{S}^{(\alpha)} \cdot \boldsymbol{S}^{(\beta)}. \tag{8.1}$$

Here $J(r^{(\alpha\beta)})$ is the exchange integral, a function of the distance $r^{(\alpha\beta)}$ from atom α to atom β; $\boldsymbol{S}^{(\alpha)}$ is the spin angular momentum, in units of $\hbar$, of atom α. We assume that $\boldsymbol{S}^{(\alpha)}$ and $\boldsymbol{S}^{(\beta)}$, for nearest neighbors, are nearly parallel. Then if $\boldsymbol{v}^{(\alpha)}$ is a unit vector in the direction of $-\boldsymbol{S}^{(\alpha)}$ (and therefore, for electron spins, in the direction of the magnetic moment of atom α), so that $\boldsymbol{S}^{(\alpha)} = -S\boldsymbol{v}^{(\alpha)}$, and if $\vartheta^{(\alpha\beta)}$ is the small angle between the directions $\boldsymbol{v}^{(\alpha)}$ and $\boldsymbol{v}^{(\beta)}$,

$$\left.\begin{aligned} W^{(\alpha\beta)} &= -2J(r^{(\alpha\beta)})\, S^2 \cos\vartheta^{(\alpha\beta)} \\ &= -2J(r^{(\alpha\beta)})\, S^2 \left(1 - \tfrac{1}{2}\vartheta^{(\alpha\beta)\,2}\right) \\ &= -2J(r^{(\alpha\beta)})\, S^2 + J(r^{(\alpha\beta)})\, S^2 (\boldsymbol{v}^{(\beta)} - \boldsymbol{v}^{(\alpha)})^2, \end{aligned}\right\} \tag{8.2}$$

since for small $\vartheta^{(\alpha\beta)}$, $|\vartheta^{(\alpha\beta)}| = |\boldsymbol{v}^{(\beta)} - \boldsymbol{v}^{(\alpha)}|$. We assume that $\boldsymbol{v}^{(\alpha)}$ can be fitted to a continuous function $\boldsymbol{v}(X_A)$ of position in the undeformed lattice and that, to a sufficient approximation,

$$\boldsymbol{v}^{(\beta)} - \boldsymbol{v}^{(\alpha)} = (\boldsymbol{R}^{(\beta)} - \boldsymbol{R}^{(\alpha)}) \cdot \nabla_0 \boldsymbol{v}, \tag{8.3}$$

where ∇_0 has components $\partial/\partial X_A$. Since the components of $\boldsymbol{v}$ are the direction cosines α_i, Eq. (8.3) may be written

$$\alpha_i^{(\beta)} - \alpha_i^{(\alpha)} = (X_A^{(\beta)} - X_A^{(\alpha)})\, \alpha_{i,A}. \tag{8.4}$$

For simplicity, we take the origin at undeformed site α; then $X_A^{(\alpha)} = 0$. Substitution of Eq. (8.3) in Eq. (8.2) gives

$$\left.\begin{aligned} W^{(\alpha\beta)} = -2J(r^{(\alpha\beta)})\, S^2 + J(r^{(\alpha\beta)})\, S^2 X_A^{(\beta)} X_B^{(\beta)}\, \alpha_{i,A}\, \alpha_{i,B} \\ \text{(not summed over } \beta). \end{aligned}\right\} \tag{8.5}$$

To find the energy per unit undeformed volume, $\varrho_0 \mathsf{F}_{\text{ex}}$, we sum Eq. (8.5) over nearest neighbors β, multiply by $\frac{1}{2}$ to avoid counting each term twice, and multiply by the number n_0 of atoms per unit undeformed volume. The contribution from the first term in Eq. (8.5) depends only on the strains E_{AB} and not on the direction-cosine gradients $\alpha_{i,A}$; it may therefore be omitted here and, instead, absorbed into the part of $\varrho_0 \mathsf{F}$ (to be considered later) dependent on E_{AB} and α_i but not on $\alpha_{i,A}$. Thus we get

$$\varrho_0 \mathsf{F}_{\text{ex}} = \tfrac{1}{2} n_0\, S^2 \sum_\beta J(r^{(\alpha\beta)})\, X_A^{(\beta)} X_B^{(\beta)}\, \alpha_{i,A}\, \alpha_{i,B} \tag{8.6}$$

(we indicate sums over β explicitly by Σ_β; for sums over i, A, etc., we use the summation convention).

If R is the undeformed distance between an atom and one of its nearest neighbors, then since the distances R and $r^{(\alpha\beta)}$ are very small on a continuum scale, we have to a sufficient approximation [cf. Eqs. (3.13), (3.14)]

$$\left.\begin{aligned} r^{(\alpha\beta)2} &= x_p^{(\alpha\beta)}\, x_p^{(\alpha\beta)} = x_{p,A}\, x_{p,B}\, X_A^{(\beta)} X_B^{(\beta)} \\ &= C_{AB} X_A^{(\beta)} X_B^{(\beta)} = X_A^{(\beta)} X_B^{(\beta)} + (C_{AB} - \delta_{AB}) X_A^{(\beta)} X_B^{(\beta)} \\ &= R^2 + 2E_{AB} X_A^{(\beta)} X_B^{(\beta)} = R^2\{1 + 2E_{AB}\, \Gamma_A^{(\beta)} \Gamma_B^{(\beta)}\}, \end{aligned}\right\} \tag{8.7}$$

where $\Gamma_A^{(\beta)} \equiv X_A^{(\beta)}/R$ are the direction cosines of the line from atom α to atom β in their undeformed positions. Hence to the first order in E_{AB}

$$r^{(\alpha\beta)} = R\{1 + E_{AB}\, \Gamma_A^{(\beta)} \Gamma_B^{(\beta)}\}, \tag{8.8}$$

and

$$J(r^{(\alpha\beta)}) = J(R) + J'(R)\, (r^{(\alpha\beta)} - R) = J(R) + J'(R)\, R E_{AB}\, \Gamma_A^{(\beta)} \Gamma_B^{(\beta)}, \tag{8.9}$$

where $J'(R) \equiv dJ(R)/dR$. On inserting this expression in Eq. (8.6), we get

$$\varrho_0 \mathsf{F}_{\text{ex}} = \tfrac{1}{2}\{b_{AB}\, \alpha_{i,A}\, \alpha_{i,B} + b_{ABCD}\, E_{AB}\, \alpha_{i,C}\, \alpha_{i,D}\}, \tag{8.10}$$

where

$$b_{AB} = n_0\, S^2 R^2 J(R) \sum_\beta \Gamma_A^{(\beta)} \Gamma_B^{(\beta)}, \tag{8.11}$$

$$b_{ABCD} = n_0\, S^2 R^3 J'(R) \sum_\beta \Gamma_A^{(\beta)} \Gamma_B^{(\beta)} \Gamma_C^{(\beta)} \Gamma_D^{(\beta)}. \tag{8.12}$$

Eq. (8.10) is equivalent to Eq. (7.43).

For a cubic crystal, $\sum_\beta \Gamma_A^{(\beta)} \Gamma_B^{(\beta)} = \frac{1}{3} Z \delta_{AB}$, where Z is the number of nearest neighbors; therefore $b_{AB} = \delta_{AB} C$, with

$$C = \tfrac{1}{3} n_0 Z\, S^2 R^2 J(R), \tag{8.13}$$

and the first term in Eq. (8.10) reduces to $\frac{1}{2} C \alpha_{i,A} \alpha_{i,A} = \frac{1}{2} C [(\nabla_0 \alpha_1)^2 + (\nabla_0 \alpha_2)^2 + (\nabla_0 \alpha_3)^2]$.

As a microscopic formula, Eq. (8.10) is valid only at $T=0$, where it evaluates the internal energy density $\varrho_0 u_{\text{ex}}$ due to exchange forces in a classical approximation. We may, however, reinterpret it as a phenomenological formula for the internal *free* energy density $\varrho_0 \mathsf{F}_{\text{ex}}$ at arbitrary temperature T; the parameters b_{AB} and b_{ABCD} must now be regarded as functions of T. Whether Eq. (8.10) is sufficient as it stands, or must be augmented by terms of higher order in the strains, must then be decided on the basis of the success of the formula in fitting experimental data. In any case, truncation of the series after the term linear in E_{AB} is a quite different matter from adoption of the approximations of infinitesimal strain theory; the E_{AB}'s in Eq. (8.10) are still finite strains, related to the displacement gradients $u_{i,A}$ by Eq. (3.50), not by Eq. (3.54). If higher-order terms are needed, their form can be determined without difficulty.

8.3. The anisotropy, magnetostrictive, and elastic energies. In the standard theory of a rigid ferromagnetic body, the exchange-energy expression is based directly on a microscopic model; the "anisotropy" energy, on the other hand, is based on phenomenological considerations, because no completely satisfactory microscopic model has been devised.[1] The same is true of the standard theory of magnetostriction: the theory has been more successful in fitting data to phenomenological formulas than in interpreting those formulas microscopically. We shall therefore consider the non-exchange parts of the local free energy (per unit mass) first from a phenomenological point of view.

In accordance with the discussion at the end of § 7.5, we assume [Eq. (7.44)]

$$\mathsf{F} = \mathsf{F}_{\text{ex}} + \mathscr{F}(\overline{\mathsf{M}}_P, E_{AB}), \tag{8.14}$$

---

[1] In the first theoretical work on magnetocrystalline anisotropy, by Akulov [2], a magnetic-quadrupole model was used to derive the expression $K_1 (\alpha_1^2 \alpha_2^2 + \alpha_2^2 \alpha_3^2 + \alpha_3^2 \alpha_1^2)$ for the leading term in the anisotropy energy-density of a cubic crystal. The term is now usually attributed to spin-orbit interactions.

where F_{ex} is given by Eq. (7.43) or (8.10), and where $\bar{\mathsf{M}}_P \equiv \mathsf{M}_i R_{Pi}$ are the components of **M** in axes that share the rotation R of the mass element under consideration. To make the formula more specific, we assume for $\mathscr{F}$ a series expansion in the strains,

$$\left.\begin{aligned}\mathscr{F}(\bar{\mathsf{M}}_P, E_{AB}) = g(\bar{\mathsf{M}}_P) + g_{AB}(\bar{\mathsf{M}}_P) E_{AB} + \\ + \tfrac{1}{2} g_{ABCD}(\bar{\mathsf{M}}_P) E_{AB} E_{CD} + \cdots,\end{aligned}\right\} \tag{8.15}$$

where the coefficients g are functions of the $\bar{\mathsf{M}}_P$'s. Finally, we may assume for these functions polynomials in the $\bar{\mathsf{M}}_P$'s (or in the corresponding direction cosines $\bar{\alpha}_P \equiv \bar{\mathsf{M}}_P/\mathsf{M}_s$). The result of all this will be a polynomial in the $\bar{\mathsf{M}}_P$'s and E_{AB}'s, with coefficients dependent only on the temperature; the number of independent coefficients can be considerably reduced by use of symmetry arguments, since most of the materials of interest are either cubic or hexagonal.

In Eqs. (8.10) and (8.15), we may suppose without loss of generality that $b_{AB} = b_{BA}$, $b_{ABCD} = b_{BACD} = b_{ABDC}$, $g_{AB} = g_{BA}$, and $g_{ABCD} = g_{BACD} = g_{CDAB}$.

We have in this section lumped together the "anisotropy, magnetostrictive, and elastic energies". In conventional magnetostriction theory, these names are usually applied to the terms of degrees 0, 1, and 2, respectively, in the strains. Often, however, there is a change from strains to stresses as independent variables, and the names may then be determined by the degree in the stresses rather than in the strains. The two modes of separation are not equivalent, and therefore confusion is likely if a separation of this sort is attempted at all. We shall for this reason avoid it. It is, in any case, a question only of semantics, not of physics or of mathematics.[1]

It can be argued that in Eq. (8.15), it would be better to follow Toupin and use the variables $\Pi_P \equiv \mathsf{M}_i x_{i,P}$ instead of the variables

[1] Kittel [*1*], p. 555, writes: "It is of the primary importance to recognize that *there will be no linear magnetostriction if the anisotropy energy is independent of the state of strain of the crystal.* Magnetostriction occurs because the anisotropy energy depends on the strain ..." Obviously this is a different labeling from that just described: Kittel would consider the terms linear in the strains (and perhaps also the terms quadratic in the strains, if they are magnetization-dependent) to be part of the "anisotropy" energy. Lee [*1*], pp. 198–199, adopts Kittel's point of view and adds: "This gives a physical explanation for the occurrence of magnetostriction which ... does not seem to have been stated until Kittel's formulation." There is, of course, nothing wrong about applying the name "anisotropy energy" to all the terms that are magnetization-dependent, rather than just to those that are also strain-independent; but it is difficult to see how this lexicographical reform can give a physical explanation of anything — or how earlier writers, merely by choosing a different lexicography, could have earned Kittel's criticism: "The close physical relationship which exists between the anisotropy and magnetostriction constants is not clearly revealed in the standard discussions ..."

$\bar{\mathsf{M}}_P \equiv \mathsf{M}_i R_{Pi}$. The formal mathematics is simpler with the Π_P's; furthermore, it seems impossible to express the exchange term F_{ex} conveniently in terms of the $\bar{\mathsf{M}}_P$'s. On the other hand, a microscopic model will usually lead to the $\bar{\mathsf{M}}_P$'s rather than to the Π_P's. The $\bar{\mathsf{M}}_P$'s have a clear physical meaning as components in axes that "rotate with" the mass element; more explicitly, in local axes derived from the x_i axes, at the position of the mass element dm, by subjecting them to the finite rotation R_{Pi}. In contrast, TOUPIN's [*1*] Π_A's are so lacking in direct physical meaning that he himself remarked (p. 904): "We shall refrain from giving the vector Π_A any physical interpretation." Thus the use here of $\bar{\mathsf{M}}_P$ rather than of Π_P has no better basis than a physicist's prejudices; a mathematician will doubtless prefer Π_P.

8.4. A microscopic model for magnetostriction. In this section we shall consider in some detail a simple microscopic model used by NÉEL [*1*]; NÉEL's calculation has been summarized by KNELLER [*1*], pp. 231 ff. The model is classical and itself somewhat phenomenological: it merely postulates certain properties of the energy of interaction of neighboring atoms, without investigating the origin of this energy. The same model, applied to the *elastic* energy, leads to the CAUCHY relations (LOVE [*1*], p. 100), which are not verified experimentally. The model itself should therefore not be taken too seriously; our primary object is to illustrate the differences between an infinitesimal-strain treatment and a finite-strain treatment of the same model.

We omit the magnetic dipole-dipole energy, which will be discussed in § 8.5; we suppose that the rest of the interaction energy of two atoms falls off so fast with distance that an atom may be considered to interact only with its nearest neighbors. To obtain the energy terms in which we are interested, it will be sufficient to consider the case in which the magnetic moments of the atom under consideration and of its neighbors are all parallel.

Let r be the distance from the central atom to one of its nearest neighbors, and let φ be the angle between their line of centers and the direction of the moments. Then if the only departure from spherical symmetry of an atom is that associated with its dipole moment, simple symmetry considerations show that the interaction energy of the two atoms is of the form

$$w = g_1(r)\, P_2(\cos\varphi) + g_2(r)\, P_4(\cos\varphi) + \cdots, \tag{8.16}$$

where the P_n's are the Legendre polynomials. [The function $g_1(r)$ does not include the magnetic dipole-dipole contribution, $-(\gamma/4\pi)\cdot(2\mu^2/r^3)$, where μ is the magnetic moment of an atom; if this contribution were included, it would not be permissible to limit consideration to nearest neighbors.]

To derive the leading magnetostriction terms, it is sufficient to consider the first term in Eq. (8.16). We write it

$$w = f(r)\,(\cos^2\varphi - \tfrac{1}{3}), \tag{8.17}$$

where $f(r) = \frac{3}{2} g_1(r)$. We choose as reference state a state of zero deformation, with M_A equal to the value of $\overline{\mathsf{M}}_A$ in the actual state under consideration; that is, we suppose that during the deformation, **M** remains fixed with respect to axes that rotate with the mass element. Then if $\bar{\alpha}_A \equiv \overline{\mathsf{M}}_A/\mathsf{M}_s$, the $\bar{\alpha}_A$'s remain constant during the change from the reference state to the actual state. In the reference state, $r = R$ and $\varphi = \Phi$; in the change to the actual state, the first-order change in w is

$$\Delta w = (\cos^2\Phi - \tfrac{1}{3})\,f'(R)\,\Delta r + f(R)\,\Delta(\cos^2\varphi), \tag{8.18}$$

where $\Delta r \equiv r - R$ and $\Delta(\cos^2\varphi) \equiv \cos^2\varphi - \cos^2\Phi$. By Eq. (8.8), to the first order in the strains E_{AB},

$$\Delta r = R\,\Gamma_A\,\Gamma_B\,E_{AB}, \tag{8.19}$$

where Γ_A are the direction cosines of the line from the central atom to the nearest neighbor under consideration, in the reference state. We must derive the analogous formula for $\Delta(\cos^2\varphi)$.

We start with the relation

$$r\cos\varphi = \alpha_i\,\Delta x_i, \tag{8.20}$$

where Δx_i are the components of the vector from the central atom to its neighbor, and where α_i are the direction cosines of **M**, both in the deformed state. Now to a sufficient approximation, $\Delta x_i = x_{i,A}\,X_A$ (we take the origin of X_A at the central atom); and $\alpha_i = \bar{\alpha}_C\,R_{Ci}$. Hence

$$r\cos\varphi = x_{i,A}\,R_{Ci}\,\bar{\alpha}_C\,X_A. \tag{8.21}$$

Postmultiplication of Eq. (3.19) with R_{Ci} (followed by the implied summation) gives

$$x_{i,A}\,R_{Ci} = (C^{\frac{1}{2}})_{AB}\,R_{Bi}\,R_{Ci} = (C^{\frac{1}{2}})_{AB}\,\delta_{BC} = (C^{\frac{1}{2}})_{AC}, \tag{8.22}$$

so that Eq. (8.21) becomes

$$r\cos\varphi = (C^{\frac{1}{2}})_{AC}\,\bar{\alpha}_C\,X_A, \tag{8.23}$$

and

$$r^2\cos^2\varphi = (C^{\frac{1}{2}})_{AC}\,(C^{\frac{1}{2}})_{BD}\,\bar{\alpha}_C\,\bar{\alpha}_D\,X_A\,X_B. \tag{8.24}$$

By Eq. (8.7)

$$r^2 = C_{AB}\,X_A\,X_B. \tag{8.25}$$

Hence

$$\cos^2\varphi = P/Q, \tag{8.26}$$

where

$$P = (C^{\frac{1}{2}})_{AC}\,(C^{\frac{1}{2}})_{BD}\,\bar{\alpha}_C\,\bar{\alpha}_D\,X_A\,X_B \tag{8.27}$$

and

$$Q = C_{EF} X_E X_F. \tag{8.28}$$

In the reference state, $C_{AB} = (C^{\frac{1}{2}})_{AB} = \delta_{AB}$, $P = \bar{\alpha}_A \bar{\alpha}_B X_A X_B = R^2 \bar{\alpha}_A \bar{\alpha}_B \Gamma_A \Gamma_B$, $Q = X_E X_E = R^2$, and Eq. (8.26) reduces to

$$\cos^2 \Phi = \Gamma_A \Gamma_B \bar{\alpha}_A \bar{\alpha}_B. \tag{8.29}$$

By expanding P/Q to the first order in E_{AB}, we find

$$\Delta(\cos^2 \varphi) = 2[\Gamma_A \bar{\alpha}_B \Gamma_C \bar{\alpha}_C - \bar{\alpha}_C \Gamma_C \bar{\alpha}_D \Gamma_D \Gamma_A \Gamma_B] E_{AB}. \tag{8.30}$$

To find the free energy per unit undistorted volume, we substitute Eqs. (8.19) and (8.30) in Eq. (8.18), sum over nearest neighbors, and multiply by $\frac{1}{2} n_0$, where n_0 is the number of atoms per unit undeformed volume. This gives an expression of the form of the term $g_{AB}(\overline{\mathsf{M}}_P) E_{AB}$ in Eq. (8.15). We need not concern ourselves with the details of this calculation; our main concern is with the relation of Eqs. (8.19) and (8.30) to the corresponding equations of NÉEL's theory.

NÉEL, using infinitesimal-strain theory, assumes that $\mathbf{M}$ has a constant direction α_i and that the lattice undergoes a strain described by the symmetric infinitesimal-strain tensor A_{ij}. His formulas are equivalent to Eqs. (8.19) and (8.30) with $\bar{\alpha}_1, \bar{\alpha}_2, \bar{\alpha}_3$ replaced by $\alpha_1, \alpha_2, \alpha_3$ and with $E_{11}, E_{12}, \ldots$ replaced by $A_{11}, A_{12}, \ldots$. Since $\bar{\alpha}_1 = \alpha_1$ to the zeroth order in $u_{i,A}$, the distinction between them is unimportant in a coefficient of a strain in a formula valid only to the first order in the displacement gradients. Thus our finite-strain calculation gives a result that reduces to NÉEL's in the limiting case to which NÉEL's calculation applies. The difference is that the present calculation can be extended to any order in the strains, by merely carrying out the expansions to correspondingly higher orders.

8.5. The magnetic self-energy. The rigorous microscopic formula for the magnetic interaction energy of a system of dipole moments $\boldsymbol{\mu}_p$, located at positions $\boldsymbol{r}_p$, is

$$W_\mu = -\tfrac{1}{2} \sum_p \boldsymbol{\mu}_p \cdot \boldsymbol{h}_p, \tag{8.31}$$

where $\boldsymbol{h}_p$ is the magnetic field intensity, at the position of dipole p, due to all the other dipoles. By Eq. (2.12),

$$\boldsymbol{h}_p = \frac{\gamma}{4\pi} \sum_q{}' (-\boldsymbol{\mu}_q + \boldsymbol{\mu}_q \cdot \mathbf{1}_{qp} \mathbf{1}_{qp}) r_{qp}^{-3}. \tag{8.32}$$

Here

$$\boldsymbol{r}_{qp} = \boldsymbol{r}_p - \boldsymbol{r}_q \quad \text{and} \quad \mathbf{1}_{qp} = \boldsymbol{r}_{qp}/r_{qp}.$$

For a rigid magnetized body, we can derive a continuum approximation to Eq. (8.31) by the well-known method of LORENTZ [1] (p. 137).

About dipole p, construct a sphere of radius R, large in comparison with interatomic distances but small in comparison with specimen dimensions. Then the sum over dipoles $\boldsymbol{\mu}_q$ outside the sphere R can be replaced, with negligible error, by an integral over moment elements $\boldsymbol{M}\,d\tau$. This integral can be transformed to an expression that is formally the field intensity of poles of volume density $-\nabla\cdot\boldsymbol{M}$ in the part of the specimen outside the Lorentz sphere, poles of surface density M_n on the external surface of the specimen, and poles of surface density $-M_n$ (where the positive normal is radially outward) on the surface of the sphere. This expression differs from $\boldsymbol{H}_1\equiv\boldsymbol{H}-\boldsymbol{H}_0$, the part of the magnetizing force $\boldsymbol{H}$ that is due to the magnetization of the specimen, by including the field of the surface charges on the sphere and by not including the field of volume charges inside it. On the supposition that the magnetization varies no worse than linearly over the sphere, the latter term is zero and the former is $\frac{1}{3}\gamma\boldsymbol{M}$. Thus

$$\boldsymbol{h}_p=\boldsymbol{H}_1+\tfrac{1}{3}\gamma\,\boldsymbol{M}+\boldsymbol{h}_p^*, \tag{8.33}$$

where $\boldsymbol{h}_p^*$ is the field intensity of dipoles (other than p) inside the Lorentz sphere. On substituting Eq. (8.33) in Eq. (8.31) and replacing the sum over p by an integration, we get

$$\left.\begin{aligned} W_\mu&=-\tfrac{1}{2}\int\boldsymbol{M}\cdot(\boldsymbol{H}_1+\tfrac{1}{3}\gamma\,\boldsymbol{M})\,d\tau+W_{\mu l}\\ &=W_{\rm m}-\tfrac{1}{6}\gamma\int\boldsymbol{M}^2\,d\tau+W_{\mu l}\end{aligned}\right\} \tag{8.34}$$

where

$$W_{\mu l}=-\tfrac{1}{2}\sum_p\boldsymbol{\mu}_p\cdot\boldsymbol{h}_p^*. \tag{8.35}$$

The value of $\boldsymbol{\mu}_p\cdot\boldsymbol{h}_p^*$ is determined by dipole moments in a "physically small" region about dipole p; a sum of this quantity over p may therefore be replaced by the integral over τ of an energy density dependent only on local conditions. The integral $\int\boldsymbol{M}^2\,d\tau$ in (8.34) is already in this form. Therefore both the second and the third terms in Eq. (8.34) may be omitted from the dipole-dipole energy and absorbed, instead, into the local free-energy density; the term $W_{\rm m}$ in W_μ is all that needs to be retained in order to take proper account of the long-range property of dipole-dipole interactions. In this respect, $W_{\rm m}$ is not unique; we might have used, instead, $W_{\rm m}'\equiv-\frac{1}{2}\int\boldsymbol{M}\cdot\boldsymbol{B}_1\,d\tau=W_{\rm m}-\frac{1}{2}\gamma\int M^2\,d\tau$ or some linear combination $\alpha W_{\rm m}+(1-\alpha)W_{\rm m}'$ $(0\leqq\alpha\leqq 1)$.

Thus for a rigid body

$$W_\mu=W_{\rm m}+\int f_{\rm m}\,d\tau, \tag{8.36}$$

where $f_{\rm m}$ is a free-energy density dependent on local conditions. In a phenomenological theory, $f_{\rm m}$ may be lumped together with the rest of the non-exchange part of the free-energy density; there is no need to consider it separately; symmetry considerations and thermodynamic arguments can be applied to the whole free-energy density. Only in a complete microscopic theory would explicit attention to f_{m}, as a distinct term in the free-energy density, be necessary.

We must now extend this calculation to a deformable body. If we still surround dipole p by a sphere, we have the difficulty that dipoles q pass from inside to outside the sphere, or *vice versa*, as the deformation changes. Instead, therefore, we surround dipole p by a surface that encloses a constant group of material particles and that becomes a Lorentz sphere in the undeformed state. Since the surface is very small on a megascopic scale, we may treat the deformation as uniform (*i.e.*, the deformation gradients $x_{i,A}$ as constant) over it; therefore the Lorentz sphere deforms to an ellipsoid. [If we measure X_A and x_i from the position of dipole p, the equation of the Lorentz sphere, $X_A X_A = \text{const} = R^2$, becomes, in x_i-space, $X_{A,i} X_{A,j} x_i x_j = R^2$, or $c_{ij} x_i x_j = R^2$, where $c_{ij} = X_{A,i} X_{A,j} = \text{const}$; since this remains a closed surface, it must be an ellipsoid.]

Instead of Eq. (8.33), we now have

$$\boldsymbol{h}_p = \boldsymbol{H}_1 + \tfrac{1}{3}\gamma \boldsymbol{\Lambda} \cdot \boldsymbol{M} + \boldsymbol{h}_p^*, \tag{8.37}$$

where $\boldsymbol{\Lambda}$ is a strain-dependent tensor with trace unity and with principal axes along the principal axes of the Lorentz ellipsoid. Instead of Eq. (8.34) we have

$$W_\mu = W_{\mathrm{m}} - \tfrac{1}{6}\gamma \int \boldsymbol{M} \cdot \boldsymbol{\Lambda} \cdot \boldsymbol{M}\, d\tau + W_{\mu l}. \tag{8.38}$$

The term W_{m} still takes account of the long-range character of the dipole-dipole energy; the other terms in W_μ may still be expressed as volume integrals of a free-energy density dependent only on local conditions; but now they are strain-dependent and therefore not as simple in form as before. They can nevertheless be absorbed into the non-exchange part of the local free energy, the mass integral of the term $\mathscr{F}$ in Eq. (8.14).

This justifies our use of W_{m} as "magnetic self-energy". It is not the whole of the dipole-dipole energy; but the difference between it and that whole is the integral over the mass elements of a local free-energy per unit mass, and the latter can be absorbed into the energy expressions considered in § 8.3. Only in a complete microscopic theory would it need to be evaluated separately.

9. The Small-Displacement Approximation

9.1. A first-order approximation. We now seek an approximation that suitably generalizes the approximations of conventional elasticity theory, discussed in § 3.4.

The basic mechanical formulas to be approximated are the equilibrium equations (7.35), (7.36), with F given by Eqs. (8.10) and (8.14). We write the equilibrium equations, in accordance with § 7.6, in the form

$$t_{ij,j}+M_j H_{i,j}+\varrho f_i=0 \quad \text{in } V, \tag{9.1}$$

$$t_{ij} n_j-\tfrac{1}{2}\gamma M_{\mathrm{n}}^2 n_i-T_i=0 \quad \text{on } S, \tag{9.2}$$

with

$$t_{ij}=\varrho \frac{\partial \mathsf{F}}{\partial x_{i,A}} x_{j,A}. \tag{9.3}$$

We write F in the form

$$\mathsf{F}=\mathsf{F}_{\mathrm{ex}}+\mathscr{F}(\overline{\mathsf{M}}_p, E_{AB}), \tag{9.4}$$

with

$$\mathsf{F}_{\mathrm{ex}}=(2\varrho_0 \mathsf{M}_{\mathrm{s}}^2)^{-1}\{b_{AB}\,\mathsf{M}_{i,A}\,\mathsf{M}_{i,B}+b_{ABCD}\,E_{AB}\,\mathsf{M}_{i,C}\,\mathsf{M}_{i,D}\} \tag{9.5}$$

and

$$\mathscr{F}(\overline{\mathsf{M}}_P, E_{AB})=g(\overline{\mathsf{M}}_P)+g_{AB}(\overline{\mathsf{M}}_P)E_{AB}+\tfrac{1}{2}g_{ABCD}(\overline{\mathsf{M}}_P)E_{AB}E_{CD}+\cdots. \tag{9.6}$$

Without loss of generality, we suppose that $b_{AB}=b_{BA}$, $b_{ABCD}=b_{BACD}=b_{ABDC}$, $g_{AB}=g_{BA}$, and $g_{ABCD}=g_{BACD}=g_{CDAB}$.

The obvious procedure for generalizing the approximations of elasticity theory is to expand F to the second order in the displacement gradients $u_{i,A}$; then t_{ij}, calculated by Eq. (9.3) (in which $\partial/\partial x_{i,A}$ may be replaced by $\partial/\partial u_{i,A}$), will be correct to the second order. In Eqs. (9.1), (9.2), consistent approximations must be made in the magnetic terms $M_j H_{i,j}$ and $-\frac{1}{2}\gamma M_{\mathrm{n}}^2 n_i$ and in the factor ϱ; to do this, we must express the magnetic terms in G (namely, W_{m} and $-\int \boldsymbol{M}\cdot\boldsymbol{H}_0\,d\tau$) to the second order in the displacement $\boldsymbol{u}$. Correctness of Eq. (9.6) to the second order in $u_{i,A}$ requires that E_{AB} in the second term be expressed to the second order in the $u_{i,A}$'s [*i.e.*, by Eq. (3.50)].

This procedure would be laborious and would apparently lead to very complicated formulas. For example, when there are no exchange forces, the components of the tensor t_{ij} in the rotated $d\overline{X}_A$ axes are in this approximation[1,2]

$$\left.\begin{aligned}\bar{t}_{AB}=\overline{M}_{[A}\overline{H}_{B]}+\tfrac{1}{2}\{\overline{M}_{[B}\overline{H}_{P]}E_{PA}+\overline{M}_{[A}\overline{H}_{P]}E_{PB}\}+\\ +\varrho\{g_{AB}+g_{BP}E_{PA}+g_{AP}E_{PB}+g_{ABPQ}E_{PQ}\};\end{aligned}\right\} \tag{9.7}$$

this expression must still be transformed to unrotated axes, in such a way that the result is correct to the first order in the $u_{i,A}$'s. The expansion of the magnetic energy terms to the second order in $\boldsymbol{u}$ would also be difficult and tedious.

---

[1] The symbol $\bar{t}_{AB}$ used here must not be confused with the symbol $\bar{t}_{ij}$ defined by Eqs. (5.26).

[2] The derivation of this formula will not be given, since the derivation is tedious and since we shall make no further use of the formula.

The procedure indicated might be necessary for some purposes; but for consideration of magnetostriction in ferromagnetic materials, it is unnecessarily complicated in application, tho simple in principle. It fails to take account of the smallness of magnetostrictive strains (of order 10^{-5}) and consequently includes terms that are actually not significant.

To illustrate this, consider a simple system for which $G = kE + \frac{1}{2} c E^2 - E f$, where E is a (finite) strain, f a measure of applied force, and k and c constants. Suppose also that E is given, in terms of displacement gradients, by a formula similar to Eq. (3.50). In equilibrium at given f, $\partial G / \partial E = 0$, whence $E = (f - k)/c$; and when $f = 0$, $E = -k/c$. If the displacement gradients are small quantities of order ε thruout the range of interest, so too is E; and if this is to be true thruout a range that includes $f = 0$, then k/c must be a small quantity of order ε. Then in the term kE in G, E is multiplied by a first-order small quantity; and it itself needs to be expressed only to the first order, if G as a whole is to be expressed consistently to the second order.

Without trying to generalize this argument, we shall proceed directly to the approximation that it suggests. Most of the rest of this monograph will use that approximation. It must be emphasized, however, that the approximation is not necessarily sufficient for all cases. In fact, EASTMAN [*1, 2*] has shown that in some cases third-order terms are experimentally significant. In such cases it is of course necessary to return to the finite-strain theory and to derive appropriate working formulas directly from it.

9.2. A more drastic approximation. We assume that the only part of G in which terms of second order in u_i or $u_{i,A}$ are significant is the part due to the term $\frac{1}{2} g_{ABCD}(\overline{M}_P) E_{AB} E_{CD}$ in Eq. (9.6); and that in all other terms dependent both on magnetization and on displacements, it is sufficient to keep the terms that are linear in u_i or $u_{i,A}$.[1] Regarding the term $\frac{1}{2} g_{ABCD}(\overline{M}_P) E_{AB} E_{CD}$, we assume that terms of higher than second order may be omitted; then this term may be replaced by $\frac{1}{2} g_{ABCD}(M_P)$

[1] We require this not only of the internal part $\int F\, dm$ of G, but also of the terms $-\int \varrho_0 f_i x_i d\tau_0 - \int T_i' x_i d S_0$. The latter requirement is satisfied if $\varrho_0 f_i$ (the force per unit undistorted volume) and T_i' (the force per unit undistorted area) are constant thruout the deformation. It is also satisfied if the forces f_i and T_i' do not exceed, in order of magnitude, those necessary to undo the magnetostrictive strains; then it is immaterial whether we write T_i' or T_i. In our applications, it will be supposed that $f_i = 0$ and that T_i does not exceed, in order of magnitude, 'T'$_i$ of Eq. (10.12), *i.e.* the negative of the surface tractions necessary to undo the magnetostrictive strains. We can therefore write T_i' or T_i indiscrimately. In a measurement of YOUNG's modulus, T_i' may exceed 'T'$_i$ in order of magnitude. The notation T_i' should then be retained; constancy of T_i' is a good approximation in this case.

$u_{(A,B)}\, u_{(C,D)}$, with argument M_P instead of $\bar{\mathsf{M}}_P$ and with the first-order approximation $u_{(A,B)}$ to E_{AB}.

In the term $g_{AB}(\bar{\mathsf{M}}_P)\, E_{AB}$, we may, by virtue of our assumptions, replace E_{AB} by $u_{(A,B)}$ and the argument $\bar{\mathsf{M}}_P$ by the argument M_P; this term therefore becomes $g_{AB}(\mathsf{M}_P)\, u_{(A,B)}$. In the term $g(\bar{\mathsf{M}}_P)$, however, we must retain terms of order 1 in $u_{i,A}$. To this order, by Eq. (3.53),

$$\bar{\mathsf{M}}_P = \mathsf{M}_i\, R_{Pi} = \mathsf{M}_i\,(\delta_{Pi} + u_{[i,P]}) = \mathsf{M}_P + \mathsf{M}_i\, u_{[i,P]}, \tag{9.8}$$

and therefore

$$g(\bar{\mathsf{M}}_P) = g(\mathsf{M}_P) + \frac{\partial g(\mathsf{M}_P)}{\partial \mathsf{M}_Q}\, \mathsf{M}_i\, u_{[i,Q]}. \tag{9.9}$$

In Eq. (9.5), we may replace E_{AB} by $u_{(A,B)}$.

With all these substitutions, Eqs. (9.4) to (9.6) give

$$\left.\begin{aligned} \mathsf{F} = {} & (2\varrho_0\, \mathsf{M}_s^2)^{-1}\{b_{AB}\, \mathsf{M}_{i,A}\, \mathsf{M}_{i,B} + b_{ABCD}\, u_{(A,B)}\, \mathsf{M}_{i,C}\, \mathsf{M}_{i,D}\} + \\ & + g(\mathsf{M}_P) + \frac{\partial g(\mathsf{M}_P)}{\partial \mathsf{M}_Q}\, \mathsf{M}_i\, u_{[i,Q]} + \\ & + g_{AB}(\mathsf{M}_P)\, u_{(A,B)} + \frac{1}{2}\, g_{ABCD}(\mathsf{M}_P)\, u_{(A,B)}\, u_{(C,D)}. \end{aligned}\right\} \tag{9.10}$$

In $t_{ij} = \varrho\,(\partial \mathsf{F}/\partial x_{i,A})\, x_{j,A} = \varrho\,(\partial \mathsf{F}/\partial u_{i,A})\, x_{j,A}$, the differentiation gives a value of $\partial \mathsf{F}/\partial u_{i,A}$ valid only to the zeroth order in the small quantities $u_{i,A}$; therefore the value of t_{ij} itself is also valid only to this order, and the factors ϱ and $x_{j,A}$ differ from ϱ_0 and δ_{jA} by terms comparable with ones already neglected. Thus we may write

$$\left.\begin{aligned} t_{ij} = \varrho_0\, \partial \mathsf{F}/\partial u_{i,j} = {} & (2\mathsf{M}_s^2)^{-1}\, b_{ijkl}\, \mathsf{M}_{m,k}\, \mathsf{M}_{m,l} + M_{[i}\, \frac{\partial g}{\partial \mathsf{M}_{j]}} + \\ & + G_{ij} + c_{ijkl}\, u_{(k,l)}, \end{aligned}\right\} \tag{9.11}$$

with[1]

$$G_{ij} = \varrho_0\, g_{ij}, \qquad c_{ijkl} = \varrho_0\, g_{ijkl}. \tag{9.12}$$

In Eq. (9.11), $M_i = \varrho_0\, \mathsf{M}_i$ in the present approximation. Since first-order small quantities are being neglected in t_{ij}, the terms that contain M or ϱ in the equilibrium equations (9.1) and (9.2) need to be evaluated to the zeroth order only, *i.e.* at $\boldsymbol{u} = 0$.

In the zeroth order, by Eq. (9.10),

$$\frac{\partial \mathsf{F}}{\partial \mathsf{M}_{j,B}} = (\varrho_0\, \mathsf{M}_s^2)^{-1}\, b_{AB}\, \mathsf{M}_{j,A} \tag{9.13}$$

and

$$\frac{\partial \mathsf{F}}{\partial \mathsf{M}_j} = \frac{\partial g}{\partial \bar{\mathsf{M}}_j}. \tag{9.14}$$

---

[1] We write c_{ijkl} for $\varrho_0\, g_{ijkl}$ in order to conform with the usual notation c for an elastic constant. It would then be consistent to write c_{ij} for $\varrho_0\, g_{ij}$. We use G_{ij} instead in order to reserve the symbol c_{ij} for an elastic constant in the abbreviated Voigt notation, $c_{11} = c_{1111}$ etc. [see Eq. (10.20)].

The magnetic equilibrium equation (7.28) therefore gives, in zeroth order,

$$M_s^{-2} b_{AB} M_{[i} M_{j],AB} - M_{[i} \frac{\partial g}{\partial \mathsf{M}_{j]}} + M_{[i} H_{j]} = 0. \tag{9.15}$$

By Eqs. (9.11) and (9.15), the antisymmetric part of t_{ij} is

$$t_{[ij]} = M_{[i} \frac{\partial g}{\partial \mathsf{M}_{j]}} = M_{[i} H_{j]} + M_s^{-2} b_{AB} M_{[i} M_{j],AB}. \tag{9.16}$$

With cubic symmetry, $b_{AB} = C\,\delta_{AB}$, this becomes

$$t_{[ij]} = M_{[i} H_{j]} + \frac{C}{M_s^2} M_{[i} \nabla^2 M_{j]}. \tag{9.17}$$

Exchange forces, as well as magnetic forces, produce a couple proportional to the size of the element it acts upon and therefore contribute to asymmetry of the stresses.

The symmetric part of t_{ij} is

$$t_{(ij)} = (2M_s^2)^{-1} b_{ijkl} M_{m,k} M_{m,l} + G_{ij} + c_{ijkl} u_{(k,l)}. \tag{9.18}$$

The mechanical equilibrium equations are Eqs. (9.1), (9.2), with

$$t_{ij} = t_{(ij)} + t_{[ij]}; \tag{9.19}$$

$t_{(ij)}$ and $t_{[ij]}$ are given by Eqs. (9.18) and (9.16) respectively, and all the magnetic quantities are evaluated at $\boldsymbol{u} = 0$. The magnetic equilibrium equations at $\boldsymbol{u} = 0$ are Eqs. (2.61), (2.62). Only these are needed for use with the mechanical equilibrium equations; but if magnetic equilibrium equations valid to the first order in u_i and $u_{i,A}$ are desired, they can be obtained by expansion of the general magnetic equilibrium equations (7.24), (7.25).

9.3. Verification by a variational method. In the last section, we began with the equilibrium equations — which result from setting the first variation of the thermodynamic potential G equal to zero — and introduced in them the approximations described at the beginning of the section. The procedure is somewhat complicated and inelegant. The same results can be obtained more simply and elegantly by first introducing the approximations in the formula for G and then carrying out the variational process on this approximate expression.

The general expression for G is given by Eq. (7.13), which may be written

$$\left. \begin{aligned} G = \int \varrho_0 \mathsf{F}\, d\tau_0 + W_m - \int \varrho_0 \boldsymbol{H}_0 \cdot \mathbf{M}\, d\tau_0 - \\ - \int \varrho_0 f_i x_i\, d\tau_0 - \int T_i' x_i\, dS_0. \end{aligned} \right\} \tag{9.20}$$

In accordance with the approximations described at the beginning of § 9.2, we are to replace F, in the first term of G, by the approximate expression (9.10), and to replace all the other terms in G by approximations valid to the first order in u_i and $u_{i,A}$. The only term of second order in u_i and $u_{i,A}$ that then remains is the term in $\int \varrho_0 \mathsf{F}\, d\tau_0$ that is contributed by the term $\frac{1}{2} g_{ABCD}(\mathsf{M}_P)\, u_{(A,B)}\, u_{(C,D)}$ in F.

The first-order term in W_{m} can be obtained from Eq. (7.31) by replacing δx_i by u_i. Thus

$$W_{\mathrm{m}} = W_{\mathrm{m}}^{(0)} + W_{\mathrm{m}}^{(1)}, \tag{9.21}$$

where $W_{\mathrm{m}}^{(0)} = -\frac{1}{2}\int \boldsymbol{H}_1 \cdot \boldsymbol{M}\, d\tau_0$, evaluated at $\boldsymbol{u}=0$, and where

$$W_{\mathrm{m}}^{(1)} = -\int M_j H_{1i,j}\, u_i\, d\tau_0 - \tfrac{1}{2}\gamma \int M_{\mathrm{n}}^2\, n_i\, u_i\, dS_0; \tag{9.22}$$

all quantities in this expression, other than u_i itself, are to be evaluated at $u_i = 0$. Similarly, the first-order term in $W_{H0} \equiv -\int \varrho_0 \boldsymbol{H}_0 \cdot \mathbf{M}\, d\tau_0$ can be obtained from Eq. (7.32) by replacing δx_i by u_i; it is

$$W_{H0}^{(1)} = -\int M_j H_{0i,j}\, u_i\, d\tau_0, \tag{9.23}$$

with everything except u_i evaluated at $u_i = 0$. Thus

$$W_{\mathrm{m}}^{(1)} + W_{H0}^{(1)} = -\int M_j H_{i,j}\, u_i\, d\tau_0 - \tfrac{1}{2}\gamma \int M_{\mathrm{n}}^2\, n_i\, u_i\, dS_0. \tag{9.24}$$

Finally, the first-order terms in the last two integrals in Eq. (9.20) are the same integrals with x_i replaced by u_i and with all other quantities evaluated at $u_i = 0$; this implies that we may replace T_i' by T_i.

On collecting the terms of common order in u_i and $u_{i,A}$, we have

$$G = G^{(0)} + G^{(1)} + G^{(2)}, \tag{9.25}$$

where $G^{(n)}$ is of order n. We need not write $G^{(0)}$; it is the same, apart from trivial differences of notation, as G of Eq. (4.22). If we set its first variation equal to zero for arbitrary variations $\delta\mathbf{M}$ subject to $\mathbf{M}^2 = \mathsf{M}_{\mathrm{s}}^2$, we get again the magnetic equilibrium equations (2.61), (2.62) for a rigid body; in the present application, these are the magnetic equilibrium equations of a deformable body to the zeroth order in u_i and $u_{i,A}$, and in them all quantities are to be evaluated at $u_i = 0$.

For $G^{(1)}$ and $G^{(2)}$ we have

$$\left.\begin{aligned} G^{(1)} = \int \Big\{ & (2\mathsf{M}_{\mathrm{s}}^2)^{-1}\, b_{ABCD}\, u_{(A,B)}\, \mathsf{M}_{i,C}\, \mathsf{M}_{i,D} + \varrho_0 \frac{\partial g(\mathsf{M}_P)}{\partial \mathsf{M}_Q}\, \mathsf{M}_i\, u_{[i,Q]} + \\ & + \varrho_0\, g_{AB}(\mathsf{M}_P)\, u_{(A,B)} \Big\}\, d\tau_0 - \int M_j H_{i,j}\, u_i\, d\tau_0 - \\ & - \frac{1}{2}\gamma \int M_{\mathrm{n}}^2\, n_i\, u_i\, dS_0 - \int \varrho_0 f_i\, u_i\, d\tau_0 - \int T_i\, u_i\, dS_0, \end{aligned}\right\} \tag{9.26}$$

$$G^{(2)} = \int \frac{1}{2} \varrho_0\, g_{ABCD}(\mathsf{M}_P)\, u_{(A,B)}\, u_{(C,D)}\, d\tau_0. \tag{9.27}$$

In Eq. (9.26), all quantities except the u_i's and $u_{i,A}$'s are to be evaluated at $u_i = 0$. The deformed coordinates x_i no longer occur as independent variables in any functions; therefore we may now abandon the distinction between capital and lower-case subscripts. We shall in the remaining steps use lower-case with the understanding that $_{,i}$ now means $\partial/\partial X_i$.

To derive the mechanical equilibrium equations, we set $\delta G^{(1)}+\delta G^{(2)}=0$ for arbitrary variations $\delta u_i(X_A)$.

To find $\delta G^{(1)}$, we replace u by δu in Eq. (9.26). At the same time, we change to lower-case subscripts; and to facilitate the necessary transformations, we also make the following minor changes. First, we remove the symmetrization parentheses from the $u_{(i,j)}$'s; this is permissible because of the symmetry properties imposed on the coefficients [see the statement after Eq. (9.6)]. Second, we shift the antisymmetrization brackets in the second term from $u_{i,Q}$ to the other factor; this is permissible because, for any tensor t_{ijkl}, $t_{ij[ij]}=\frac{1}{2}(t_{ijij}-t_{ijji})=\frac{1}{2}(t_{ijij}-t_{jiij})=t_{[ij]ij}$. Thus we get

$$\left.\begin{aligned}\delta G^{(1)}=\int\Big\{&(2\mathsf{M}_s^2)^{-1}\,b_{ijkl}\,\mathsf{M}_{m,k}\,\mathsf{M}_{m,l}+\varrho_0\frac{\partial g(\mathsf{M}_k)}{\partial \mathsf{M}_{[j}}\,\mathsf{M}_{i]}+\\&+\varrho_0\,g_{ij}(\mathsf{M}_k)\Big\}\,\delta u_{i,j}\,d\tau_0-\int(M_j\,H_{i,j}+\varrho_0\,f_i)\,\delta u_i\,d\tau_0-\\&-\int\Big(\frac{1}{2}\gamma\,M_n^2\,n_i+T_i\Big)\,\delta u_i\,dS_0.\end{aligned}\right\}\tag{9.28}$$

For $\delta G^{(2)}$, we easily find

$$\delta G^{(2)}=\int\varrho_0\,g_{ijkl}(\mathsf{M}_m)\,u_{k,l}\,\delta u_{i,j}\,d\tau_0.\tag{9.29}$$

In $\delta G^{(1)}+\delta G^{(2)}$, let the symbol t_{ij} represent the coefficient of $\delta u_{i,j}$, namely

$$\left.\begin{aligned}t_{ij}\equiv\,&(2\mathsf{M}_s^2)^{-1}\,b_{ijkl}\,\mathsf{M}_{m,k}\,\mathsf{M}_{m,l}+\varrho_0\,\mathsf{M}_{[i}\frac{\partial g(\mathsf{M}_k)}{\partial\mathsf{M}_{j]}}+\\&+\varrho_0\,g_{ij}(\mathsf{M}_k)+\varrho_0\,g_{ijkl}\,(\mathsf{M}_m)\,u_{(k,l)};\end{aligned}\right\}\tag{9.30}$$

insertion of the symmetrization parentheses in the last term is justified by the symmetry properties of g_{ijkl}. Then we have, for variations of u_i,

$$\left.\begin{aligned}\delta G=\delta G^{(1)}+\delta G^{(2)}=&\int\{t_{ij}\,\delta u_{i,j}-(M_j\,H_{i,j}+\varrho_0\,f_i)\,\delta u_i\}\,d\tau_0-\\&-\int(\tfrac{1}{2}\gamma\,M_n^2\,n_i+T_i)\,\delta u_i\,dS_0.\end{aligned}\right\}\tag{9.31}$$

In this equation, all quantities except the $u_{k,l}$'s, δu_i's, and $\delta u_{i,j}$'s are to be evaluated at $u_i=0$; the $u_{k,l}$'s occur only in the last term of t_{ij} and are definite functions, tho unknown ones (until the equilibrium equations have been solved), of the undeformed coordinates (X_1, X_2, X_3) over which we are integrating; the δu_i's are arbitrary functions of these same coordinates, and the $\delta u_{i,j}$'s are their derivatives with respect to these coordinates; the change from capital to lower-case subscripts was a matter of formal notation, possible because the deformed coordinates x_i no longer appear as independent variables in any of our functions. It follows that we may transform the integral $\int t_{ij}\,\delta u_{i,j}\,dS_0$ in the usual way, in a space in which the coordinates are (X_1, X_2, X_3) and $_{,j}$ means $\partial/\partial X_j$:

$$\int t_{ij}\,\delta u_{i,j}\,d\tau_0 = -\int t_{ij,j}\,\delta u_i\,d\tau_0 + \int t_{ij}\,n_j\,\delta u_i\,dS_0. \tag{9.32}$$

On substituting the expression (9.32) for the first term in Eq. (9.31) and equating to zero the coefficients of δu_i in V_0 and on S_0, we get the equilibrium equations

$$t_{ij,j} + M_j\,H_{i,j} + \varrho_0\,f_i = 0 \quad \text{in } V_0, \tag{9.33}$$

$$t_{ij}\,n_j - \tfrac{1}{2}\gamma\,M_{\mathrm{n}}^2\,n_i - T_i = 0 \quad \text{on } S_0. \tag{9.34}$$

Here t_{ij} is given by Eq. (9.30); the displacement u_i occurs only in the last term of t_{ij}; and all other quantities, both in Eq. (9.30) and in Eqs. (9.33), (9.34), are to be evaluated in the undeformed state, $u_i = 0$. It is easily seen that these results are equivalent to those of § 9.2.

Magnetic equilibrium equations correct to the first (or second) order in u_i and $u_{i,A}$ can be obtained by varying $G^{(0)} + G^{(1)}$ (or $G^{(0)} + G^{(1)} + G^{(2)}$) with respect to $\mathbf{M}$, under the constraint $\mathbf{M}^2 = \mathrm{M}_{\mathrm{s}}^2$, at constant u_i. The calculation is straightforward but will not be carried out here, since the results are at present of no great interest: even the zero-order magnetic equilibrium equations have so far defied solution except in trivial cases, and therefore there is not much point in extending them to the first or second order in the strains. In § 11 we shall consider the combined magnetic and mechanical equilibrium equations in a special case in which the equations have been linearized (with respect to $\mathbf{M}$) and are therefore, in principle, solvable.

9.4. Comparison with conventional magnetostriction theory. The conventional theory of magnetostriction, described in § 1.2, makes all the approximations of § 9.2 plus the following additional ones: in $G^{(2)}$ of Eq. (9.27), $\varrho_0\,g_{ABCD} = c_{ABCD}$ is assumed to be independent of $\mathbf{M}$, and the first, second, fourth, and fifth terms in $G^{(1)}$ of Eq. (9.26) are omitted. Thus in this theory

$$G^{(1)} = \int [G_{ij}(\mathrm{M}_m)\,u_{i,j} - \varrho_0\,f_i\,u_i]\,d\tau_0 - \int T_i\,u_i\,dS_0, \tag{9.35}$$

$$G^{(2)} = \int \tfrac{1}{2}\,c_{ijkl}\,u_{i,j}\,u_{k,l}\,d\tau_0, \tag{9.36}$$

where $G_{ij}(\mathrm{M}_m) = G_{ji}(\mathrm{M}_m)$, and where $c_{ijkl} = c_{klij} = c_{jikl}$ is independent of $\mathbf{M}$. Use of Eqs. (9.35), (9.36) is equivalent to the assumption that the Helmholtz function F is the sum of three terms: the F of a rigid magnetizable body, the F of a nonmagnetic body, and an interaction term of the form $\int G_{ij}(\mathrm{M}_m)\,u_{(i,j)}\,d\tau_0$.

The first and second terms in Eq. (9.26) are responsible for the first and second terms in t_{ij} of Eq. (9.30); in other words, for the first (exchange) term in the symmetric part of t_{ij}, Eq. (9.18), and for the whole antisymmetric part, given by Eq. (9.16). Omission of these terms reduces t_{ij} to

$$t_{ij} = G_{ij} + c_{ijkl}\, u_{(k,l)}. \tag{9.37}$$

The fourth and fifth terms in Eq. (9.26) are responsible for the term $M_j H_{i,j}$ in the volume equilibrium equation (9.33) and the term $-\frac{1}{2}\gamma M_n^2 n_i$ in the surface equilibrium equation (9.34). These therefore become

$$t_{ij,j} + \varrho_0 f_i = 0 \quad \text{in } V_0, \tag{9.38}$$

$$t_{ij} n_j - T_i = 0 \quad \text{on } S_0. \tag{9.39}$$

As compared with the conventional theory, summarized in Eqs. (9.37) to (9.39), the improved theory (still with the approximations of § 9.2) adds to t_{ij} the antisymmetric term $M_{[i} H_{j]}$ due to magnetic couples, another antisymmetric term due to exchange couples, and a symmetric term due to the dependence of exchange forces on strains [see Eqs. (9.16) and (9.18)]. In the equilibrium equations, it adds an effective volume force density $M_j H_{i,j}$ and an effective surface force density $\frac{1}{2}\gamma M_n^2 n_i$.[1]

In a uniformly magnetized ellipsoid in a uniform applied field, with $\boldsymbol{M}$ along $\boldsymbol{H}$, all these additions vanish except the surface force density $\frac{1}{2}\gamma M_n^2 n_i$. The theory then reduces to the theory of the nonuniform form effect in an ellipsoid; this subject will be treated in § 10.

9.5. Uniqueness of the strains. Suppose that the following quantities are given: the applied field $\boldsymbol{H}_0$; the magnetic moment per unit mass $\mathbf{M}$, and hence $\boldsymbol{H}$, to the zeroth order in u_i and $u_{i,j}$; and the body and surface forces f_i and T_i. The values of f_i and T_i must satisfy the conditions (7.79) and (7.82) for the given $\boldsymbol{H}_0$ and $\mathbf{M}$. Under these specifications, let $u_i^{(1)}$ and $u_i^{(2)}$ be two solutions of the partial differential equations (9.33), with t_{ij} given by Eq. (9.30); and let $u_i = u_i^{(2)} - u_i^{(1)}$ be the difference between the two solutions. Then u_i satisfies the partial differential equation

$$t_{ij,j} = 0, \tag{9.40}$$

where

$$t_{ij} \equiv t_{ij}^{(2)} - t_{ij}^{(1)} = \varrho_0\, g_{ijkl}(\mathsf{M}_m)\, u_{(k,l)}, \tag{9.41}$$

and the boundary condition

$$t_{ij} n_j = 0. \tag{9.42}$$

---

[1] These statements are dependent on our arbitrary choice of t_{ij}, now defined by Eq. (9.30), as the generalization of the "stress" tensor of a nonmagnetic body. We could instead have used t'_{ij}, $\bar{t}_{ij}$, $\bar{t}'_{ij}$, or τ_{ij} of § 5.6 for this purpose. Our statements would then be different; but all five sets of statements become equivalent if we eliminate the auxiliary quantities t_{ij}, t'_{ij}, $\bar{t}_{ij}$, $\bar{t}'_{ij}$, or τ_{ij} and describe directly the total effect of the improved theory on the equilibrium equations (in the form that results from the elimination).

These, however, are precisely the equations satisfied by u_i in a non-magnetic elastic body with elastic constants c_{ijkl} and with zero applied forces f_i and T_i. The uniqueness theorem of standard elasticity theory (LOVE [1], pp. 170—171) therefore applies and shows that $u_{(i,j)}=0$. (The theorem holds if the quadratic form $\frac{1}{2}c_{ijkl}\,u_{(i,j)}\,u_{(k,l)}$ is positive definite; it must be, if the state of strain at fixed $\boldsymbol{M}$ is to be stable.) In other words, the sets of strains $u^{(1)}_{(i,j)}$ and $u^{(2)}_{(i,j)}$ are identical, and the displacements $u^{(1)}_i$ and $u^{(2)}_i$ can differ at most by a rigid-body displacement and rotation. Thus the solution for the strains at given $\mathbf{M}$ is unique.

Concerning the uniqueness of the solution for $\mathbf{M}$ at given $\boldsymbol{u}$, or of the solution for $\boldsymbol{M}$ and $\boldsymbol{u}$ jointly, no general theoretical statement can be made, because of the nonlinearity of the magnetic equilibrium equations. We shall later (§ 11) examine linearized cases in which the solution changes from uniqueness to nonuniqueness as the applied field H_0 is changed. On physical grounds, nonuniqueness of the solution must occur in the range of field intensities in which, experimentally, hysteresis is observed. This is illustrated by the case of the rigid ellipsoidal single-domain particle (STONER and WOHLFARTH [1]), in which the stable equilibrium states (one for sufficiently large $|H_0|$, two for sufficiently small) can be calculated exactly.

Chapter IV

Applications

10. The Magnetostriction of a Uniformly Magnetized Ellipsoid

10.1. Finite-strain formulas. In §10 we shall consider in some detail the case of a physically homogeneous body whose shape, in the undeformed state, is ellipsoidal, with the principal axes of the ellipsoid along the coordinate axes X_A. We shall suppose that the moment per unit mass $\mathbf{M}$ and the applied field $\boldsymbol{H}_0$ are uniform and along the coordinate axis X_3. We shall first suppose that the deformation gradients $x_{i,A}$ are uniform and shall determine whether the state described can be an equilibrium state under no body or surface forces; having demonstrated that it cannot, we shall then investigate, in the small-displacement approximation, the nature of the actual equilibrium strain distribution, with $\mathbf{M}$ still uniform to the zeroth order in the strains.

We can carry out the initial stages of this study by rigorous finite-strain theory. With $x_{i,A}=\text{const}\equiv A_{iA}$, the deformation consists, in general, of a rigid-body rotation R_{Ai} followed by elongations in the ratios $C_1^{\frac{1}{2}}$, $C_2^{\frac{1}{2}}$, $C_3^{\frac{1}{2}}$ along the principal axes of the tensor C_{AB}. To keep the problem reasonably simple, we shall suppose that $R_{Ai}=\delta_{Ai}$ (no rotation) and that the principal axes of C_{AB} coincide with the X_A axes. Then (if we keep the center of the body fixed and choose it as origin)

$$x_1 = C_1^{\frac{1}{2}} X_1, \qquad x_2 = C_2^{\frac{1}{2}} X_2, \qquad x_3 = C_3^{\frac{1}{2}} X_3, \tag{10.1}$$

so that the original ellipsoid

$$\frac{X_1^2}{a_1^2} + \frac{X_2^2}{a_2^2} + \frac{X_3^2}{a_3^2} = 1 \tag{10.2}$$

becomes the ellipsoid

$$\frac{x_1^2}{C_1 a_1^2} + \frac{x_2^2}{C_2 a_2^2} + \frac{x_3^2}{C_3 a_3^2} = 1. \tag{10.3}$$

In order that a state of uniform magnetization along X_3, with applied field $\boldsymbol{H}_0$ in the same direction, may be a state of magnetic equilibrium, the specimen must satisfy certain physical and geometrical symmetry requirements; it will be sufficient if, for example, the specimen is a single crystal of cubic symmetry, with the [*100*] directions oriented along the X_A axes. There is then no torque of crystalline origin to disturb the magnetic equilibrium. The total magnetizing force is along the magnetization and equal to $H_0 - \gamma D_3 M$, where D_3 is the demagnetizing factor for magnetization along the X_3 axis (Brown [*10*], p. 33); the couple density $\boldsymbol{M} \times \boldsymbol{H}$ is therefore zero. The magnetic equilibrium will be stable if H_0 is sufficiently large.

Since the $x_{i,A}$'s are constant with respect to the X_A's, $J = \det(x_{i,A})$ and $\varrho = \varrho_0/J$ are constant. Since M_i is also constant and $\mathsf{M}_{i,A}$ is zero, $t_{ij} = \varrho[\partial \mathsf{F}(\mathsf{M}_k, x_{k,P}, \mathsf{M}_{k,P})/\partial x_{i,A}]\, x_{j,A}$, a function of the variables M_k, $x_{k,P}$, and $\mathsf{M}_{k,P}$, is constant. Since $\mathsf{M}_{j,A} = 0$, Eq. (7.53) reduces to

$$t_{[ij]} = -\varrho \frac{\partial \mathsf{F}}{\partial \mathsf{M}_{[i}} \mathsf{M}_{j]}; \tag{10.4}$$

Since $\varrho(\partial \mathsf{F}/\partial \mathsf{M}_{i,A})\, x_{j,A}$ is constant, Eq. (7.24) reduces to

$$\mathsf{M}_{[k} \left\{ \varrho \frac{\partial \mathsf{F}}{\partial \mathsf{M}_{i]}} - \varrho H_{i]} \right\} = 0. \tag{10.5}$$

From Eqs. (10.4) and (10.5),

$$t_{[ij]} = M_{[i} H_{j]} = 0, \tag{10.6}$$

since $\boldsymbol{M} \times \boldsymbol{H} = 0$.

The mechanical equilibrium equations (7.48) and (7.49) become, since $t_{ij,j}$ and $H_{i,j}$ vanish,

$$\varrho f_i = 0 \qquad \text{in } V, \tag{10.7}$$

$$T_i = t_{ij} n_j - \tfrac{1}{2} \gamma M_{\mathrm{n}}^2 n_i \quad \text{on } S. \tag{10.8}$$

Equilibrium under no applied forces ($f_i = 0$, $T_i = 0$) therefore requires that the constants C_1, C_2, C_3 be so chosen that the constants t_{ij} satisfy the condition $t_{ij} n_j - \frac{1}{2}\gamma M_{\text{n}}^2 n_i = 0$ at each point of S.

In the limiting case of an infinitely long cylinder with axis along X_3, $n_3 = 0$ and $M_{\text{n}} = 0$, and equilibrium requires $t_{i1} n_1 + t_{i2} n_2 = 0$; since this must hold both where $n_1 = 1$ ($\therefore n_2 = 0$) and where $n_2 = 1$ ($\therefore n_1 = 0$), both t_{i1} and t_{i2} must vanish. In the opposite limiting case of a plate with infinite plane faces perpendicular to X_3, $n_1 = n_2 = 0$, $n_3 = 1$, and $M_{\text{n}} = M$; equilibrium requires $t_{13} = t_{23} = 0$, $t_{33} = \frac{1}{2}\gamma M^2$. In these limiting cases, the equilibrium conditions can be satisfied with constant $x_{i,A}$'s. For a finite ellipsoid, satisfaction of the boundary conditions at the ends of the principal axes requires $t_{i1} = t_{i2} = 0$, $t_{i3} = \frac{1}{2}\gamma M^2 n_i$; then at an arbitrary point, $t_{ij} n_j - \frac{1}{2}\gamma M_{\text{n}}^2 n_i = t_{i3} n_3 - \frac{1}{2}\gamma M_{\text{n}}^2 n_i = \frac{1}{2}\gamma M^2 n_i n_3 - \frac{1}{2}\gamma M^2 n_3^2 n_i = \frac{1}{2}\gamma M^2 n_i n_3 (1 - n_3)$, which in general does not vanish. Thus the assumed state of uniform $x_{i,A}$ is, for a finite ellipsoid, not an equilibrium state.

This conclusion is independent of the form of the function F. To find the actual strain distribution, for uniform **M** along X_3, we must assume some specific function F. For this purpose, we shall now go over to the small-displacement approximation.

10.2. The small-displacement approximation. We continue to suppose that **M** and $\boldsymbol{H}$ are uniform and along X_3, at least to the zeroth order in u_i and $u_{i,A}$; but we now go over to the approximations of §§ 9.2, 9.3. Since $\mathsf{M}_{m,k} = 0$, Eq. (9.16) becomes $t_{[ij]} = M_{[i} H_{j]}$; and since $\boldsymbol{M} \times \boldsymbol{H} = 0$, $t_{[ij]} = 0$. Therefore t_{ij} is equal to its symmetric part $t_{(ij)}$; and by Eq. (9.18) with $M_{m,k} = 0$, this becomes

$$t_{ij} = G_{ij} + c_{ijkl}\, u_{(k,l)}. \tag{10.9}$$

It must be remembered that in the present approximation, magnetic terms such as $M_{[i} H_{j]}$ are to be evaluated to the zeroth order in u_i and $u_{i,A}$; that is, for the undeformed body. The quantities c_{ij} and c_{ijkl} are functions of the M_i's (or α_i's); but since $\boldsymbol{M}$ is now uniform, the G_{ij}'s and c_{ijkl}'s are constant (with respect to X_1, X_2, X_3).

With no body forces ($f_i = 0$) and with uniform $\boldsymbol{H}$ ($H_{i,j} = 0$), the volume equilibrium condition reduces to $t_{ij,j} = 0$, or (since $G_{ij,j} = 0$ and $c_{ijkl,j} = 0$)

$$c_{ijkl}\, u_{(k,l)j} = 0. \tag{10.10}$$

This is the same as for a nonmagnetic body with elastic constants c_{ijkl}. The surface equilibrium condition, under zero applied surface forces ($T_i = 0$), becomes $t_{ij} n_j - \frac{1}{2}\gamma M_{\text{n}}^2 n_i = 0$, or

$$c_{ijkl}\, u_{(k,l)}\, n_j = -G_{ij} n_j + \tfrac{1}{2}\gamma M_{\text{n}}^2 n_i. \tag{10.11}$$

This is the same as for the same nonmagnetic body under mechanical applied surface forces

$$`T'_i = -G_{ij} n_j + \tfrac{1}{2} \gamma M_n^2 n_i. \tag{10.12}$$

Since the equilibrium equations are now linear in the u_i's, we may solve separately for the two sets of applied surface forces,

$$`T'^{(1)}_i = -G_{ij} n_j \tag{10.13}$$

and

$$`T'^{(2)}_i = +\tfrac{1}{2} \gamma M_n^2 n_i, \tag{10.14}$$

and superpose the results.

The problem of a nonmagnetic body with surface forces $`T'^{(1)}_i$ can be solved[1] with uniform strains $u_{(k,l)}$. These give uniform stresses $`t'_{ij} = c_{ijkl} u_{(k,l)}$; the quantity to be set equal to $`T'^{(1)}_i$ is $`t'_{ij} n_j = c_{ijkl} u_{(k,l)} n_j$. With the value (10.13) of $`T'^{(1)}_i$ inserted, this becomes

$$c_{ijkl} u_{(k,l)} n_j = -G_{ij} n_j, \tag{10.15}$$

which requires

$$c_{ijkl} u_{(k,l)} = -G_{ij}. \tag{10.16}$$

The solution is

$$u_{(i,j)} = -s_{ijkl} G_{kl}, \tag{10.17}$$

where the s_{ijkl}'s are, in VOIGT's [*1*] terminology, the "elastic moduli" corresponding to the "elastic constants" c_{ijkl}. Specifically, if we adopt VOIGT's notation (except that we use e for x and t for X),

$$e_1 = u_{1,1}, \ldots, \qquad e_4 = u_{2,3} + u_{3,2}, \ldots, \tag{10.18}$$

$$t_1 = `t'_{11}, \ldots, \qquad t_4 = `t'_{23} = `t'_{32}, \ldots, \tag{10.19}$$

the stress-strain relation $`t'_{ij} = c_{ijkl} u_{(k,l)}$ can be written

$$t_\alpha = c_{\alpha\beta} e_\beta, \tag{10.20}$$

where the Greek subscripts are summed from 1 to 6; solution of the six simultaneous equations (10.20) for the e_β's gives

$$e_\alpha = s_{\alpha\beta} t_\beta; \tag{10.21}$$

---

[1] That this solution is unique follows from the general uniqueness theorem of linear elasticity theory (LOVE [*1*], p. 170, Sect. 118). The displacements are unique except for an arbitrary rigid-body displacement and rotation that do not affect the strains. The proof hinges on the supposition that the elastic energy density $\frac{1}{2} c_{ijkl} u_{(i,j)} u_{(k,l)} (= \frac{1}{2} c_{\alpha\beta} x_\alpha x_\beta$ in VOIGT's notation) is positive for all values of the x_α's other than $x_1 = \ldots = x_6 = 0$. That this is true follows from the stability requirement $\delta^2 F > 0$ in § 4.5 (the condition that F shall be not merely stationary with respect to the variations δu_i, but actually a minimum).

if the relations (10.21) are rewritten in the form $u_{(i,j)} = s_{ijkl}\,{}^{\prime}t^{\prime}_{kl}$ (with $s_{ijkl} = s_{klij} = s_{jikl}$), the resulting s_{ijkl}'s are the desired quantities.

The corresponding problem with surface forces $\frac{1}{2}\gamma M_n^2 n_i$ cannot be solved with uniform strains. The nonuniform strains required have been determined only for a sphere, and only for certain sets of elastic constants c_{ijkl}. It is possible, however, to find the *volume average* strains for an ellipsoid of arbitrary dimension ratio, and with elastic constants of the most general type.

The strains (10.17) are independent of the shape of the specimen and are the only magnetostrictive strains considered in the conventional theory. The strains corresponding to effective surface forces $\frac{1}{2}\gamma M_n^2 n_i$ depend on the shape of the specimen. They vanish in the limit of an infinitely long cylinder extending along the X_3 axis; at the opposite extreme, a plate with plane faces of infinite extent perpendicular to the X_3 axis, they produce uniform strains (in VOIGT's notation)

$$e_\alpha = -\, s_{\alpha 3} \cdot \tfrac{1}{2}\gamma M_n^2; \tag{10.22}$$

if the elastic constants can be approximated sufficiently by those of a nonmagnetic isotropic elastic solid, this becomes (E = YOUNG's modulus, σ = POISSON's ratio)

$$e_3 = -\tfrac{1}{2}\gamma\, M_n^2/E, \qquad e_1 = e_2 = \tfrac{1}{2}\gamma\, M_n^2\,\sigma/E, \qquad e_4 = \dots = 0. \tag{10.23}$$

For any particular material, comparison of the strains (10.22) and (10.23) with the strains (10.17) will show whether the "form effect" is appreciable or negligible. The conventional theory may be regarded as giving the strains characteristic of a long cylinder with axis along the magnetization; measurements on some other shape may require a correction before the formulas of the theory can be applied to them.

In § 10.3 we shall evaluate the volume-average form-effect strains; in § 10.4 we shall examine the special case of a sphere, in which the actual nonuniform strains can be evaluated.

10.3. The average strains. The calculation of volume-average strains is based on BETTI's reciprocity theorem (LOVE [*1*], pp. 173—176): if forces f_i and T_i produce displacements u_i and if forces f_i' and T_i' produce displacements u_i', then

$$\int \varrho f_i\, u_i'\, d\tau + \int T_i\, u_i'\, dS = \int \varrho f_i'\, u_i\, d\tau + \int T_i'\, u_i\, dS. \tag{10.24}$$

The proof is simple:

$$\left.\begin{aligned} \int \varrho f_i\, u_i'\, d\tau + \int T_i\, u_i'\, dS &= \int \varrho f_i\, u_i'\, d\tau + \int t_{ij}\, u_i'\, n_j\, dS \\ &= \int \{\varrho f_i\, u_i' + (t_{ij}\, u_i')_{,j}\}\, d\tau = \int \{(\varrho f_i + t_{ij,j})\, u_i' + t_{ij}\, u_{i,j}'\}\, d\tau \\ &= \int t_{ij}\, u_{i,j}'\, d\tau, \end{aligned}\right\} \tag{10.25}$$

by the equilibrium equation (3.5) (with $a_i=0$). Since $t_{ij}=c_{ijkl}\,u_{k,l}$, with $c_{ijkl}=c_{klij}=c_{jikl}$, the last integrand can be written $c_{ijkl}\,u'_{i,j}\,u_{k,l}=c_{klij}\,u'_{i,j}\,u_{k,l}=t'_{kl}\,u_{k,l}=t'_{ij}\,u_{i,j}$; thus primed and unprimed quantities may be interchanged in the final expression in Eq. (10.25) and therefore also in the initial expression.

For our application, it is more convenient to write the theorem in the form

$$\int \varrho f_i\,u'_i\,d\tau+\int T_i\,u'_i\,dS=\int t'_{ij}\,u_{i,j}\,d\tau. \tag{10.26}$$

Let u_i be the (unknown) displacements produced by forces T_i (with $f_i=0$), and let u'_i be displacements corresponding to uniform strains $u'_{i,j}=A_{ij}$: then by specifying $u'_i=0$ at the origin and $u'_{[i,j]}=0$ (no rotation), we get

$$u'_i=A_{ij}\,x_j \quad (A_{ij}=A_{ji}=\text{const}), \tag{10.27}$$

and therefore

$$A_{ij}\int T_i\,x_j\,dS=\int c_{ijkl}\,u'_{i,j}\,u_{k,l}\,d\tau=c_{ijkl}\,A_{ij}\int u_{k,l}\,d\tau. \tag{10.28}$$

Since this holds for arbitrary A_{11}, A_{22}, A_{33}, $A_{23}\ (=A_{32})$, $A_{31}\ (=A_{13})$, and $A_{12}\ (=A_{21})$, the coefficients of $A_{11}, \ldots, A_{23}+A_{32}, \ldots$ must vanish separately; thus

$$\int T_{(i}\,x_{j)}\,dS=c_{ijkl}\int u_{k,l}\,d\tau=V c_{ijkl}\,\overline{u_{(k,l)}}, \tag{10.29}$$

where the bar denotes the volume average. Solution for $\overline{u_{(k,l)}}$ gives

$$\overline{u_{(i,j)}}=\frac{1}{V}\,s_{ijkl}\int T_{(k}\,x_{l)}\,dS. \tag{10.30}$$

In our application, the evaluation of the volume-average strains due to '$T'^{(2)}_i$' of Eq. (10.14), we replace T_i by $\frac{1}{2}\gamma M_{\mathrm{n}}^2\,n_i=\frac{1}{2}\gamma M^2\,n_3^2\,n_i$. Thus

$$\overline{u_{(i,j)}}=\frac{\gamma M^2}{2V}\,s_{ijkl}\int n_3^2\,n_k\,x_l\,dS. \tag{10.31}$$

The symmetrizing parentheses in $n_{(k}\,x_{l)}$ have been dropped because the relation $s_{ijkl}=s_{ijlk}$ guarantees the required symmetrization.

For an isotropic material,

$$s_{1111}=\frac{1}{E}, \qquad s_{1122}=s_{1133}=-\frac{\sigma}{E}, \qquad \text{other } s_{11kl}\text{'s}=0, \tag{10.32}$$

where E is Young's modulus and σ is Poisson's ratio. Thus

$$\overline{e_1}=\overline{u_{(1,1)}}=\frac{\gamma M^2}{2VE}\int n_3^2\,[n_1\,x-\sigma(n_2\,y+n_3\,z)]\,dS. \tag{10.33}$$

Similarly

$$\overline{e_2}=\overline{u_{(2,2)}}=\frac{\gamma M^2}{2VE}\int n_3^2\,[n_2\,y-\sigma(n_1\,x+n_3\,z)]\,dS, \tag{10.34}$$

$$\overline{e_3}=\overline{u_{(3,3)}}=\frac{\gamma M^2}{2VE}\int n_3^2\,[n_3\,z-\sigma(n_1\,x+n_2\,y)]\,dS. \tag{10.35}$$

With only t_4, $e_4 = t_4/\mu$, where μ is the rigidity; or $u_{2,3} + u_{3,2} = t_4/\mu$, hence $u_{(2,3)} = t_4/2\mu = t_{23}/2\mu = t_{32}/2\mu = (t_{23} + t_{32})/4\mu$. Thus

$$s_{2323} = s_{2332} = 1/4\mu, \quad \text{other} \quad s_{23kl}\text{'s} = 0, \tag{10.36}$$

and

$$\left.\begin{aligned} \overline{e_4} &= 2\overline{u_{(2,3)}} = 2 \cdot \frac{\gamma M^2}{2V} \cdot \frac{1}{4\mu} \int n_3^2 (n_2 z + n_3 y)\, dS \\ &= \frac{\gamma M^2}{4V\mu} \int n_3^2 (n_2 z + n_3 y)\, dS. \end{aligned}\right\} \tag{10.37}$$

The only use so far made of the ellipsoidal shape is the omission of terms (in the basic equations) that would be present if $\boldsymbol{H}$ were nonuniform or not along $\boldsymbol{M}$. We may now use it to show that the integral in Eq. (10.37) is zero; this is evident because for every dS with a given $n_3^2 n_2 z$, there is another dS with exactly the opposite value, *e.g.* the dS with the same x and y but opposite z. The formulas for $\overline{e_5}$ and $\overline{e_6}$ are similar. Thus

$$\overline{e_4} = \overline{e_5} = \overline{e_6} = 0. \tag{10.38}$$

Tho the form effect has been known theoretically since 1931 (POWELL [*1*], HAYASI [*1*]) and experimentally since 1934 (KORNETZKI [*1*]), it was until 1953 (BROWN [*6*]) assumed to produce uniform strains. The method of calculating it was to constrain the strains to uniformity and, under this constraint, to minimize the thermodynamic potential G with respect to them. In G, the dependence of the magnetic self-energy W_{m} on deformation was taken into account; since this effect was neglected in the conventional theory of magnetostriction, the form-effect calculation was essentially an attempt to correct, in a special case (an ellipsoid), for an omission in the general theory.

We shall now show that the strains calculated by assuming them to be uniform are equal to the volume averages (10.31) of the actual strains.

By Eqs. (9.26) and (9.27), when $\boldsymbol{H}$ is uniform and along $\mathbf{M}$, $\mathsf{M}_{i,A} = 0$, $u_{[i,j]} = 0$, $f_i = 0$, and $T_i = 0$,

$$\left.\begin{aligned} G = \text{const} - \tfrac{1}{2}\gamma \textstyle\int M_{\text{n}}^2\, n_i\, u_i\, dS_0 + \\ + \textstyle\int (G_{ij}\, u_{i,j} + \tfrac{1}{2} c_{ijkl}\, u_{i,j}\, u_{k,l})\, d\tau_0. \end{aligned}\right\} \tag{10.39}$$

It should be noted that the external field energy $-\int \boldsymbol{M} \cdot \boldsymbol{H}_0\, d\tau = -\int \mathbf{M} \cdot \boldsymbol{H}_0\, dm$ is not changed by the deformation, since we are supposing that $\mathbf{M}$ (not $\boldsymbol{M}$) remains constant. We now constrain the displacements to be ones corresponding to uniform strain, zero rotation, and $u_i = 0$ at the origin. Then (in quantities multiplied by u_i or $u_{i,j}$, we need not distinguish the x_i's from the X_A's)

$$u_i = A_{ij}\, x_j \quad (A_{ij} = A_{ji}), \tag{10.40}$$

where A_{ij} is the constant value of $u_{(i,j)}$. Eq. (10.39) now becomes

$$\left.\begin{aligned} G = \text{const} - \tfrac{1}{2}\gamma A_{ij} \int M_{\mathrm{n}}^2 n_i x_j \, dS_0 + \\ + G_{ij} A_{ij} V + \tfrac{1}{2} c_{ijkl} A_{ij} A_{kl} V; \end{aligned}\right\} \qquad (10.41)$$

and in a variation of the A_{ij}'s,

$$\left.\begin{aligned} \delta G = -\tfrac{1}{2}\gamma\, \delta A_{ij} \int M_{\mathrm{n}}^2 n_i x_j \, dS_0 + \\ + G_{ij}\, \delta A_{ij} V + c_{ijkl}\, \delta A_{ij} A_{kl} V. \end{aligned}\right\} \qquad (10.42)$$

[In the last term: $\delta(c_{ijkl} A_{ij} A_{kl}) = c_{ijkl}(A_{ij}\,\delta A_{kl} + A_{kl}\,\delta A_{ij})$; but $c_{ijkl} A_{ij}\,\delta A_{kl} = c_{klij} A_{kl}\,\delta A_{ij} = c_{ijkl} A_{kl}\,\delta A_{ij}$, since $c_{ijkl} = c_{klij}$; therefore $\delta(c_{ijkl} A_{ij} A_{kl}) = 2c_{ijkl} A_{kl}\,\delta A_{ij}$.] For a minimum, this must vanish for arbitrary δA_{11}, δA_{22}, δA_{33}, $\delta A_{23} (=\delta A_{32})$, $\delta A_{31} (=\delta A_{13})$, and $\delta A_{12} (=\delta A_{21})$. Therefore the coefficients of δA_{11}, δA_{22}, δA_{33}, $\delta A_{23} + \delta A_{32}$, $\delta A_{31} + \delta A_{13}$, and $\delta A_{12} + \delta A_{21}$ must vanish. This gives

$$-\tfrac{1}{2}\gamma \int M_{\mathrm{n}}^2 n_{(i} x_{j)} \, dS_0 + G_{(ij)} V + c_{(ij)kl} A_{kl} V = 0 \qquad (10.43)$$

or

$$c_{ijkl} A_{(kl)} = -G_{ij} + \frac{\gamma}{2V} \int M_{\mathrm{n}}^2 n_{(i} x_{j)} \, dS_0. \qquad (10.44)$$

Solution for $A_{(ij)}$ gives

$$A_{(ij)} = -s_{ijkl} G_{kl} + \frac{\gamma}{2V} s_{ijkl} \int M_{\mathrm{n}}^2 n_{(k} x_{l)} \, dS_0 \qquad (10.45)$$

or

$$A_{ij} = -s_{ijkl} G_{kl} + \frac{\gamma M^2}{2V} s_{ijkl} \int n_3^2 n_k x_l \, dS_0. \qquad (10.46)$$

The first term is the shape-independent part of $u_{(i,j)}$, Eq. (10.17). The second term is the volume-average form effect contribution to $u_{(i,j)}$, Eq. (10.31).

The first term in Eq. (10.42),

$$\delta W_{\mathrm{m}} = -\tfrac{1}{2}\gamma M^2\, \delta A_{ij} \int n_3^2 n_i x_j \, dS_0, \qquad (10.47)$$

can also be expressed in terms of the demagnetizing factor D_3 for magnetization of the ellipsoid along z. We assume such symmetry that $A_{12} = A_{23} = A_{31} = 0$. Then since $W_{\mathrm{m}} = \tfrac{1}{2}\gamma D_3 M^2 V = \tfrac{1}{2}\gamma \mathsf{M}^2 m \varrho D_3$, and since the variation is at constant mass m and magnetic moment per unit mass **M**,

$$\left.\begin{aligned} \delta W_{\mathrm{m}} &= \frac{1}{2}\gamma \mathsf{M}^2 m (D_3\,\delta\varrho + \varrho\,\delta D_3) \\ &= \frac{1}{2}\gamma \mathsf{M}^2 m \left(\varrho_0 D_3 \frac{\delta\varrho}{\varrho_0} + \varrho_0\,\delta D_3\right) \end{aligned}\right\} \qquad (10.48)$$

(we may replace ϱ by ϱ_0 in a coefficient of a first-order small quantity). Now

$$\frac{\delta\varrho}{\varrho_0} = -\frac{\delta V}{V} = -(\delta A_{11} + \delta A_{22} + \delta A_{33}) \qquad (10.49)$$

and

$$\delta D_3 = \frac{\partial D_3}{\partial A_{11}} \delta A_{11} + \frac{\partial D_3}{\partial A_{22}} \delta A_{22} + \frac{\partial D_3}{\partial A_{33}} \delta A_{33}. \quad (10.50)$$

Therefore

$$\left.\begin{aligned} \delta W_\mathrm{m} = \frac{1}{2} \gamma \, \mathsf{M}^2 m \Big[-\varrho_0 \, D_3 (\delta A_{11} + \delta A_{22} + \delta A_{33}) + \\ + \varrho_0 \left(\frac{\partial D_3}{\partial A_{11}} \delta A_{11} + \frac{\partial D_3}{\partial A_{22}} \delta A_{22} + \frac{\partial D_3}{\partial A_{33}} \delta A_{33} \right) \Big]. \end{aligned}\right\} \quad (10.51)$$

This must be equivalent to Eq. (10.47), for arbitrary δA_{ii}'s, therefore

$$-\frac{1}{2} \gamma \, M^2 \int n_3^2 \, n_1 \, x \, d S_0 = \frac{1}{2} \gamma \, \mathsf{M}^2 \, m \varrho_0 \left(-D_3 + \frac{\partial D_3}{\partial A_{11}} \right) \quad (10.52)$$

or

$$\int n_3^2 \, n_1 \, x \, d S_0 = V_0 (D_3 - \partial D_3 / \partial A_{11}). \quad (10.53)$$

Similarly

$$\int n_3^2 \, n_2 \, y \, d S_0 = V_0 (D_3 - \partial D_3 / \partial A_{22}), \quad (10.54)$$

$$\int n_3^2 \, n_3 \, z \, d S_0 = V_0 (D_3 - \partial D_3 / \partial A_{33}). \quad (10.55)$$

For isotropic material, by Eq. (10.33),

$$\overline{e_1} = \frac{\gamma M^2}{2 V_0 E} \int n_3^2 [n_1 \, x - \sigma (n_2 \, y + n_3 \, z)] \, d S_0 \quad (10.56)$$

$$= \frac{\gamma M^2}{2E} \left\{ (1 - 2\sigma) D_3 - \left[\frac{\partial D_3}{\partial A_{11}} - \sigma \left(\frac{\partial D_3}{\partial A_{22}} + \frac{\partial D_3}{\partial A_{33}} \right) \right] \right\}. \quad (10.57)$$

Now a uniform expansion, $\delta A_{11} = \delta A_{22} = \delta A_{33}$, does not affect the dimension ratios of the ellipsoid and therefore does not change D_3; consequently,

$$\frac{\partial D_3}{\partial A_{11}} + \frac{\partial D_3}{\partial A_{22}} + \frac{\partial D_3}{\partial A_{33}} = 0, \quad (10.58)$$

and we may replace $(\partial D_3 / \partial A_{22} + \partial D_3 / \partial A_{33})$ by $-\partial D_3 / \partial A_{11}$. Hence, finally,

$$\overline{e_1} = \frac{\gamma M^2}{2E} \left\{ (1 - 2\sigma) D_3 - (1 + \sigma) \frac{\partial D_3}{\partial A_{11}} \right\} = \frac{\gamma M^2 D_3}{6k} - \frac{\gamma M^2}{4\mu} \frac{\partial D_3}{\partial A_{11}}, \quad (10.59)$$

where k is the bulk modulus and μ the rigidity; for (Love [*1*], pp. 103 to 104) $E/(1 - 2\sigma) = 3k$ and $E/(1 + \sigma) = 2\mu$. For $\overline{e_2}$ and $\overline{e_3}$, A_{11} is replaced by A_{22} and A_{33} respectively.

Calculations of the form effect under the constraint to uniform strains have usually been made on the basis of the demagnetizing-factor expression for W_m and have therefore led to formulas of the form (10.59) (see Becker and Döring [*1*], pp. 303—305; Kneller [*1*], pp. 228—229; Gersdorf [*1*], Eq. (2.7)). However, direct integration of a formula such as (10.33) is also possible (Brown [*6*], Eqs. (A 10) to (A 13)).

10.4. The strains in a sphere. To find the actual strains, we must solve the problem of a nonmagnetic elastic body, with elastic constants c_{ijkl}, under normal surface tension $\frac{1}{2}\gamma M_n^2 = \frac{1}{2}\gamma M^2 n_3^2$.

It must first be pointed out that the presence of the magnetization (along X_3) lowers the symmetry of the array of elastic constants, since directions along and perpendicular to **M** are not physically equivalent.[1] For example, if the material is isotropic with respect to its combined magnetic and elastic properties, then with respect to its elastic properties alone, when the magnetization is along X_3, it is only "transversely isotropic" (LOVE [*1*], p. 160, Sec. 110, (2)) and must be expected to have five independent elastic constants (in VOIGT'S notation, $c_{11}=c_{22}$, c_{33}, $c_{44}=c_{55}$, c_{66}, $c_{12}=c_{11}-2c_{66}$, $c_{13}=c_{23}$), instead of the two of an elastically isotropic material.[2] It is usual to neglect this lowering of symmetry and to treat iron or nickel, for example, as cubic, with elastic constants independent of the direction of the magnetization, or even as isotropic.

If we adopt the last-mentioned approximation and suppose that the c_{ijkl}'s are those of an isotropic nonmagnetic material, then our problem appears at first glance simple. Actually it is not, because of the factor n_3^2 in the boundary pressure. If the equation of the ellipsoid is $f(x,y,z)=1$, where $f(x,y,z)=x^2/a^2+y^2/b^2+z^2/c^2$, then $\boldsymbol{n}$ is in the direction of ∇f, so that

$$n_3^2 = \frac{(z/c^2)^2}{(x/a^2)^2+(y/b^2)^2+(z/c^2)^2}\,. \tag{10.60}$$

The strains due to a boundary pressure containing such a factor can no doubt be solved by one or another of the methods that have been developed for ellipsoidal elastic bodies, but the calculation would require a series expansion of some kind and would be laborious. Experts in elasticity have therefore, quite understandably, preferred to spend their

[1] This effect of the direction of the spontaneous magnetization on the *microscopic* elastic constants must not be confused with the much larger "ΔE effect", which is a dependence of the *megascopic* elastic constants of an *inhomogeneously* magnetized ferromagnetic material on its microscopic magnetization distribution. Briefly, in the demagnetized state a rearrangement of the microscopic magnetization distribution (domain structure) under the influence of an applied tension can produce an additional elongation, by reorienting microscopic magnetostrictive strain systems already present; at saturation this process is suppressed; thus the total elongation per unit tension varies with the magnetization distribution. See KNELLER [*1*], pp. 705—716. We are concerned in the present discussion with stress-strain relations in a *homogeneously* magnetized region.

[2] MASON [*1*] gives formulas for a cubic crystal; he calls the deviation of the elastic constants from symmetry a "morphic" effect. He also gives some numerical values, based on wave-velocity measurements in a nickel crystal, of the effect of magnetization reorientation on certain of the elastic constants. It amounts to a fraction of 1%.

time on more rewarding projects, such as writing general treatises on the subject.

For a sphere, $a=b=c$, $n_3^2=z^2/a^2$, and the difficulties disappear. The surface force density $\boldsymbol{T}$ is radially outward and of magnitude

$$T_r=\tfrac{1}{2}\gamma M^2\cos^2\vartheta, \tag{10.61}$$

where (r, ϑ, φ) are spherical coordinates related in the usual manner to the Cartesian coordinates (X_1, X_2, X_3) or (x, y, z). The displacements and strains corresponding to radial surface traction $\varepsilon' S_n$, where ε' is a constant and S_n a surface spherical harmonic, can be found for an isotropic material by the method outlined by LOVE [*1*], p. 251, Sec. 173, (ii). The surface traction (10.61) can be written

$$T_r=\tfrac{1}{6}\gamma M^2\{P_0(\cos\vartheta)+2P_2(\cos\vartheta)\}. \tag{10.62}$$

The traction $\frac{1}{2}\gamma M^2 P_0(\cos\vartheta)=\frac{1}{2}\gamma M^2$ produces a uniform dilatation

$$\Delta=\frac{1}{6}\frac{\gamma M^2}{k}, \tag{10.63}$$

where k is the bulk modulus; or

$$\left.\begin{aligned} &e_{11}=e_{22}=e_{33}=\frac{1}{3}\Delta=\frac{1}{18}\frac{\gamma M^2}{k},\\ &e_{12}=e_{23}=e_{31}=0. \end{aligned}\right\} \tag{10.64}$$

Our notation here is

$$\left.\begin{aligned} &e_{11}\equiv u_{1,1}=u_{(1,1)}=e_1\ (\text{VOIGT}),\ \ldots,\\ &e_{23}\equiv u_{2,3}+u_{3,2}=2u_{(2,3)}=e_4\ (\text{VOIGT}),\ \ldots. \end{aligned}\right\} \tag{10.65}$$

The displacements (made unique by specifying $\boldsymbol{u}=0$ at $\boldsymbol{r}=0$ and zero rigid-body rotation) are

$$u_1=\frac{1}{18}\frac{\gamma M^2}{k}x,\ \ldots. \tag{10.66}$$

To the strains (10.64) we must add those produced by the traction $\frac{1}{3}\gamma M^2 P_2(\cos\vartheta)$. The calculation by the cited method is tedious but straightforward. The displacement due to traction $\varepsilon' P_n(\cos\vartheta)$ ($\varepsilon'=$ const) is

$$\boldsymbol{u}_n=r^2\nabla\omega_n+\alpha_n\boldsymbol{r}\,\omega_n+\nabla\varphi_n, \tag{10.67}$$

where

$$\varphi_n=-\frac{2n+\alpha_n}{2(n-1)}a^2\omega_n, \tag{10.68}$$

$$\omega_n=\frac{\varepsilon' U_n}{2n(\lambda+\mu)+[(n+3)\lambda+(n+2)\mu]\alpha_n}, \tag{10.69}$$

$$U_n=\frac{r^n}{\alpha^n}P_n(\cos\vartheta), \tag{10.70}$$

$$\alpha_n=-2\frac{n\lambda+(3n+1)\mu}{(n+3)\lambda+(n+5)\mu}; \tag{10.71}$$

a is the radius of the sphere, and λ and μ are the usual elastic constants (μ = rigidity, $k=\lambda+\frac{2}{3}\mu$ = bulk modulus). Here we set $n=2$, find ω_2 and φ_2 in spherical coordinates, rewrite them in Cartesian coordinates, and calculate the Cartesian components by Eq. (10.67). The results are

$$\left.\begin{aligned} u_1 &= -A\,x + B\{(-1-\tfrac{1}{2}\alpha)\,x^3 + (-1-\tfrac{1}{2}\alpha)\,x y^2 + (-1+\alpha)\,x z^2\},\\ u_2 &= -A\,y + B\{(-1-\tfrac{1}{2}\alpha)\,y x^2 + (-1-\tfrac{1}{2}\alpha)\,y^3 + (-1+\alpha)\,y z^2\},\\ u_3 &= 2A\,z + B\{(2-\tfrac{1}{2}\alpha)\,z x^2 + (2-\tfrac{1}{2}\alpha)\,z y^2 + (2+\alpha)\,z^3\}, \end{aligned}\right\} \quad (10.72)$$

where $\alpha=\alpha_2$ and

$$A=\varepsilon'\frac{8\lambda+7\mu}{2\mu(19\lambda+14\mu)}, \qquad B=-\frac{\varepsilon'}{a^2}\,\frac{5\lambda+7\mu}{2\mu(19\lambda+14\mu)}. \quad (10.73)$$

Differentiation of Eqs. (10.72) gives for the strains (since $\varepsilon'=\frac{1}{3}\gamma M^2$)

$$\left.\begin{aligned} e_{11} &= 3K\left\{-\frac{8\lambda+7\mu}{3} + 3\lambda\frac{x^2}{a^2} + \lambda\frac{y^2}{a^2} + (3\lambda+7\mu)\frac{z^2}{a^2}\right\},\\ e_{22} &= 3K\left\{-\frac{8\lambda+7\mu}{3} + \lambda\frac{x^2}{a^2} + 3\lambda\frac{y^2}{a^2} + (3\lambda+7\mu)\frac{z^2}{a^2}\right\},\\ e_{33} &= 3K\left\{\frac{2(8\lambda+7\mu)}{3} - (4\lambda+7\mu)\frac{x^2}{a^2} - (4\lambda+7\mu)\frac{y^2}{a^2} - 6\lambda\frac{z^2}{a^2}\right\}, \end{aligned}\right\} \quad (10.74)$$

where

$$K=\frac{\gamma M^2}{6\mu(19\lambda+14\mu)}; \quad (10.75)$$

$$e_{23}=-K'\,yz, \qquad e_{31}=-K'\,xz, \qquad e_{12}=2K'\,xy, \quad (10.76)$$

where

$$K'=\gamma M^2\frac{\lambda}{\mu(19\lambda+14\mu)}\,\frac{1}{a^2}. \quad (10.77)$$

The total strains are found by adding the partial strains (10.64) and (10.74)—(10.77). The results are equivalent to those obtained by GERSDORF [1], Eqs. (6.8), by a method that does not require evaluation of the displacements. He also found the strains for a sphere with cubic symmetry and, approximately, for some limiting forms of ellipsoid.

The volume-average strains for the sphere are

$$\left.\begin{aligned} \overline{e_{11}} &= \overline{e_{22}} = \gamma M^2\left\{\frac{1}{18k} - \frac{1}{30\mu}\right\},\\ \overline{e_{33}} &= \gamma M^2\left\{\frac{1}{18k} + \frac{1}{15\mu}\right\},\\ \overline{e_{23}} &= \overline{e_{31}} = \overline{e_{12}} = 0. \end{aligned}\right\} \quad (10.78)$$

These can be found either by averaging of the strains (10.64) and (10.74)—(10.77) or by use of Eq. (10.59) and its cyclically permuted companions (for a sphere, $D_3=\frac{1}{3}$, $\partial D_3/\partial A_{11}=\partial D_3/\partial A_{22}=\frac{2}{15}$, $\partial D_3/\partial A_{33}=-\frac{4}{15}$).

11. Problems of Micromagnetics

11.1. The distribution problem in general. The uniform magnetization assumed in § 10 is rarely attained under experimental conditions, despite the strong tendency of exchange forces to aline neighboring spins parallel. Also important are the long-range dipole-dipole interactions, described phenomenologically by the magnetic "self-energy" (see § 2.6)

$$W_{\mathrm{m}} = -\tfrac{1}{2}\int \boldsymbol{M}\cdot\boldsymbol{H}_1\, d\tau; \tag{11.1}$$

their tendency is to produce a poleless distribution, *i.e.* one for which $\nabla\cdot\boldsymbol{M}=0$ inside the specimen and $\boldsymbol{n}\cdot\boldsymbol{M}=0$ on its surface. This is obvious from the alternative expression for W_{m} [Eq. (2.50)],

$$W_{\mathrm{m}} = \frac{1}{2\gamma}\int_{\text{space}} \boldsymbol{H}_1^2\, d\tau, \tag{11.2}$$

which shows that W_{m} attains its minimum value 0 when $\boldsymbol{H}_1=0$ everywhere, *i.e.* when there are no poles. [The volume and surface pole densities $-\nabla\cdot\boldsymbol{M}$ and $\boldsymbol{n}\cdot\boldsymbol{M}$ are equal, respectively, to $+\gamma^{-1}\nabla\cdot\boldsymbol{H}_1$ and $\gamma^{-1}\boldsymbol{n}\cdot(\boldsymbol{H}_1^{+}-\boldsymbol{H}_1^{-})$; clearly these must vanish if $\boldsymbol{H}_1$ is to vanish.] In any finite specimen, uniform magnetization produces surface poles.

If we look only at one atom and one of its nearest neighbors, we find that the dipole-dipole energy, of order μ^2/R^3 (μ = moment, R = interatomic distance), is very small in comparison with the exchange energy, of order $2J(R)S^2$ [see § 8.2]; but if we examine the interaction of the first atom with atoms more distant from it, we find that the exchange energy is negligible except for nearest and perhaps next-nearest neighbors, whereas the dipole energy decreases only as $1/R^3$. This is so slow a decrease that as the size of the specimen becomes infinite, the total energy of interaction of an atom with all the other atoms, tho remaining finite, continues to depend on the shape of the specimen, even in the limit. [In the notation of § 8.5, this energy is $-\boldsymbol{\mu}\cdot\boldsymbol{h}_p$, where $\boldsymbol{h}_p=\boldsymbol{H}_1+\frac{1}{3}\gamma\boldsymbol{M}+\boldsymbol{h}_p^*$, Eq. (8.33); $\frac{1}{3}\gamma\boldsymbol{M}$ and $\boldsymbol{h}_p^*$ depend only on local conditions, but $\boldsymbol{H}_1$ depends on the magnetization distribution of the whole specimen; for a uniformly magnetized ellipsoid, $\boldsymbol{H}_1$ depends on the demagnetizing factors and thus on the axis ratios of the ellipsoid.]

The two requirements — parallel alinement of nearest-neighbor spins, and avoidance of poles — cannot both be satisfied exactly, except in an infinitely long cylinder. They can, however, both be satisfied approximately, by a magnetization distribution in which nearest-neighbor spins are almost parallel, yet over megascopic distances the direction of magnetization changes so as to keep the surface pole density $\boldsymbol{n}\cdot\boldsymbol{M}$ small on all parts of the surface, without thereby developing a large volume pole density $-\nabla \cdot \boldsymbol{M}$. This compromise between the demands of exchange forces and of dipole forces is possible only in sufficiently large bodies; in fine particles (*e.g.* iron spheres of radii less than about 2×10^{-6} cm), the spatial rate of variation necessary to avoid poles would entail excess exchange energy greater than the dipole-dipole energy avoided, and so the magnetization remains uniform.

To transform the foregoing discussion into a precise quantitative theory would require solution of the magnetic equilibrium equations (7.28), (7.29), together with the magnetostatic equations (2.27) and (2.31) (with $\boldsymbol{K}_c = 0$) that relate $\boldsymbol{H} \equiv \boldsymbol{H}_0 + \boldsymbol{H}_1$ to its sources $-\nabla \cdot \boldsymbol{M}$ and $\boldsymbol{n} \cdot \boldsymbol{M}$, and with the mechanical equilibrium equations (7.35), (7.36). After solution of this nonlinear set of equations, the stability must be tested, *i.e.* it must be shown that $\delta^2 G > 0$ for arbitrary variations $\delta \mathsf{M}_i$ (subject to $\mathsf{M}_i \mathsf{M}_i = \mathsf{M}_s^2$) and δu_i. To carry out such a calculation, it would be necessary to introduce a definite formula for the internal free-energy density F, *e.g.* Eq. (8.14) with F_{ex} given by Eq. (8.10) and $\mathscr{F}$ by Eq. (8.15). Nothing resembling such a calculation has apparently been attempted in the published literature.

For a rigid body, $u_i = 0$, the system of equations to be solved includes only the magnetic equilibrium set (7.28), (7.29) and the magnetostatic equations that relate $\boldsymbol{H}_1$ to its sources. Except in the case of fine particles, where $\boldsymbol{M}$ may be assumed to be at least approximately uniform, the calculation seems prohibitively difficult. The chief difficulties are the nonlinearity of the magnetic equilibrium equations and the necessity for solving the magnetostatic equations for each distribution $\boldsymbol{M}(x, y, z)$ considered. When $\boldsymbol{M}$ is constrained to vary only with a single Cartesian coordinate, $\boldsymbol{M} = \boldsymbol{M}(z)$ (with specimen surfaces $z = z_1$ and $z = z_2$ compatible with the constraint), the magnetostatic problem becomes trivial, and the equilibrium equations can be solved for certain forms of the anisotropy energy $g(\boldsymbol{\mathsf{M}})$; but it has been shown (Brown and Shtrikman [*1*]) that all such solutions, other than the uniform one $\boldsymbol{M} = \text{const}$, are unstable. The more realistic two- and three-dimensional problems have, in the nonlinear range, been solved only by very crude approximate methods. These usually start from the assumption, supported by experimental observation of surface distributions, that the magnetization remains nearly uniform within each of a large number of three-dimensional regions, called "domains", and that adjacent, differently magnetized domains are separated by relatively thin transition regions, "Bloch walls", within which the magnetization variation is rapid, nearly one-dimensional, and along the normal to the wall. Details of this "domain theory" have been reviewed by several authors (Brown [*1*], Kittel [*1*], Stewart [*1*], Kittel and Galt [*1*], Craik and Tebble [*1, 2*], Träuble [*1*]) and need not be discussed here.

The one case in which the nonlinear three-dimensional magnetic equilibrium equations can be solved rigorously (for a rigid specimen) is the case of the particle so fine that its magnetization may be assumed to be uniform (the "single-domain particle"). Because of the constraint to uniformity, the equilibrium equations are no longer partial differential equations but only algebraic or trigonometric equations; the simplest cases can be solved analytically (STONER and WOHLFARTH [*1*]), less simple cases numerically (WOHLFARTH [*1*, *2*]; BROWN [*11*], Chap. 6). The magnetostatic self-energy can be expressed very simply in terms of demagnetizing factors, not only for an ellipsoid but for any *uniformly magnetized* body (BROWN and MORRISH [*1*]).

Inclusion of magnetostriction in the theory of the single-domain particle requires an extension of the theory of § 10 to the case in which $\boldsymbol{M}$ is along an arbitrary direction, not necessarily the direction of $\boldsymbol{H}_0$ or of $\boldsymbol{H}$. Since $\boldsymbol{M} \times \boldsymbol{H}$ then does not in general vanish, t_{ij} is in general not symmetric. The strains may be expected to be nonuniform, as they were found to be in § 10. Solution for the nonuniform strains involves the same difficulties as in § 10. In an initial attack on the problem, the strains may be constrained to uniformity; by analogy with § 10.3, the results may be expected to give the correct average strains, tho probably not the correct strain-dependent energy. The mechanical boundary conditions imposed must be consistent with the presence of a couple $\boldsymbol{M} \times \boldsymbol{H}_0 V$ due to the applied field; equilibrium, when $\boldsymbol{M}$ is not along $\boldsymbol{H}_0$, cannot be maintained by zero tractions T_i but requires tractions that produce an equilibrium couple $-\boldsymbol{M} \times \boldsymbol{H}_0 V$. The case in which the surface displacements u_i rather than the surface forces T_i are specified involves no such difficulty.

Such a calculation for the elastically deformable single-domain particle would provide an interesting application of the general theory developed in this monograph. Since it promises to be laborious and involves some difficult questions about boundary conditions, we shall not attempt it here.

One nonlinear case that has been solved numerically, for a rigid body, is the case of an infinite cylinder with magnetization originally along the positive cylinder axis, but now in the process of reversing its magnetization by "magnetization curling"; that is, in cylindrical coordinates (ϱ, φ, z), $M_\varrho = 0$, $M_\varphi = M_s \sin\varepsilon$, $M_z = M_s \cos\varepsilon$, with $\varepsilon = \varepsilon(\varrho)$. Because $\nabla \cdot \boldsymbol{M}$ and $\boldsymbol{n} \cdot \boldsymbol{M}$ are zero, the complication of magnetostatic-energy calculation is not present. When crystalline anisotropy favors magnetization along the cylinder axis, the results (BROWN [*7*], [*11*] (§ 6.4), AHARONI and SHTRIKMAN [*1*]) show that reversal of magnetization occurs in a single discontinuous jump; one can evaluate rigorously the field at which this occurs. When the anisotropy favors magnetization perpendicular to the cylinder axis (MULLER and WEHLAU [1]), stable states of nonuniform magnetization may exist. Extension of these calculations to a deformable body might lead to equations solvable by numerical methods (by use of a fast computer), but they are of limited interest because they concern a rather special case.

Another case in which the rigid-body equations have yielded solutions is the case in which $\boldsymbol{M}$ deviates only slightly from a particular direction of minimum anisotropy-plus-external-field energy. It is then legitimate to express all the terms in F to the second order, only, in the small transverse components of $\boldsymbol{M}$ (or the corresponding direction cosines) (BROWN [3], [4], [11] (Chap. 5); SEEGER and KRONMÜLLER [1]; KRONMÜLLER [1]). The equilibrium equations, derived from F by differentiation, are therefore linear in these transverse components. The calculation is still very laborious, if the magnetostatic self-energy is taken into account; but the linearity of the equations makes the calculation possible in many cases, tho sometimes only by numerical calculations with high-speed computers.

The extension of these linearizable calculations to a deformable body will be investigated in § 11.3.

In domain theory, magnetostriction and stresses have always played an important role; the methods of taking them into account, however, have usually been very crude. Not only have they been based on the conventional theory, criticized in detail in § 9.4, but even within the framework of that theory the methods used have been inexact. The strains $u_{(i,j)}$ have usually been found by minimizing the local free-energy density $c_{ij}(\mathsf{M}_m)\,u_{(i,j)}+\frac{1}{2}c_{ijkl}\,u_{(i,j)}\,u_{(k,l)}$ with respect to the six $u_{(i,j)}$'s as independent variables. Then $c_{ij}(\mathsf{M}_m)+c_{ijkl}\,u_{(k,l)}=0$, whence $u_{(i,j)}=-s_{ijkl}\,c_{kl}(\mathsf{M}_m)$; the first of these [see Eq. (9.37)] is equivalent to $t_{ij}=0$. This is correct only when $\mathbf{M}$ and therefore $c_{ij}(\mathbf{M})$ are uniform, so that the equilibrium equations can be satisfied with uniform strains; otherwise the treatment of the $u_{(i,j)}$'s as independent variables ignores the compatibility conditions (LOVE [1], pp. 48—50). The correct equilibrium conditions are, of course, not $t_{ij}=0$ (6 equations, since in the conventional theory $t_{ij}=t_{ji}$) but $t_{ij,j}=0$ (three equations).

The justification for such approximations is that domain theory rests on very crude approximations to begin with; therefore the complicated calculations necessary to solve the correct equilibrium equations of the traditional magnetostriction theory would not be justified. Even less justification exists for including the refinements of an *improved* magnetostriction theory. Thus it seems doubtful that the present monograph has much to contribute to domain theory (it may, however, contribute something to the theory of the structure of the interdomain wall).

11.2. Cubic crystals. In any application of the theory to specific cases, it will be necessary to replace the sets of coefficients b_{AB}, b_{ABCD}, etc. by the more restrictive sets appropriate to the particular type of crystal under consideration. In this section we shall illustrate this procedure by the case of cubic symmetry, the commonest encountered in ferromagnetic materials. We take the directions of the cube edges as Cartesian axes.

Eq. (8.14), with F_{ex} and $\mathscr{F}$ from Eqs. (8.10) and (8.15), gives

$$\left.\begin{aligned}\varrho_0 \mathsf{F} = (2\mathsf{M}_s^2)^{-1}\{(b_{AB} + b_{CDAB} E_{CD}) \mathsf{M}_{i,A} \mathsf{M}_{i,B}\} + \varrho_0 g(\overline{\mathsf{M}}_P) + \\ + \varrho_0 g_{AB}(\overline{\mathsf{M}}_P) E_{AB} + \tfrac{1}{2} g_{ABCD} E_{AB} E_{CD}.\end{aligned}\right\} \tag{11.3}$$

We shall limit ourselves to the small-displacement approximation, in which Eq. (8.5) becomes Eq. (9.10); or with $G(\mathsf{M}_p) \equiv \varrho_0 g(\mathsf{M}_p)$,

$$\left.\begin{aligned}\varrho_0 \mathsf{F} = \frac{1}{2}\{(b_{ij} + b_{klij} u_{(k,l)}) \alpha_{m,i} \alpha_{m,j}\} + G(\alpha_m) + \\ + \frac{\partial G(\alpha_m)}{\partial \alpha_j} \alpha_i u_{[i,j]} + G_{ij}(\alpha_m) u_{i,j} + \tfrac{1}{2} c_{ijkl}(\alpha_m) u_{i,j} u_{k,l}.\end{aligned}\right\} \tag{11.4}$$

We have replaced M_i by $\mathsf{M}_s \alpha_i$, where α_i are the direction cosines of **M**.

We shall neglect the variation of the "elastic constants" c_{ijkl} with magnetization direction. Then the c_{ijkl}'s are those of a nonmagnetic cubic crystal. In the abbreviated Voigt notation ($e_1 = u_{1,1} = u_{(1,1)}, \ldots,$ $e_4 = u_{2,3} + u_{3,2} = 2u_{(2,3)}, \ldots$),

$$\left.\begin{aligned}\tfrac{1}{2} c_{ijkl} u_{i,j} u_{k,l} = \tfrac{1}{2} c_{11}(e_1^2 + e_2^2 + e_3^2) + c_{12}(e_2 e_3 + e_3 e_1 + e_1 e_2) + \\ + \tfrac{1}{2} c_{44}(e_4^2 + e_5^2 + e_6^2).\end{aligned}\right\} \tag{11.5}$$

Since this can be written

$$\left.\begin{aligned}\tfrac{1}{2} c_{ijkl} u_{i,j} u_{k,l} = \tfrac{1}{2} c_{11}[u_{1,1}^2 + u_{2,2}^2 + u_{3,3}^2] + \\ + \tfrac{1}{2} c_{12}[(u_{2,2} u_{3,3} + u_{3,3} u_{1,1} + u_{1,1} u_{2,2}) + \\ + (u_{3,3} u_{2,2} + u_{1,1} u_{3,3} + u_{2,2} u_{1,1})] + \\ + \tfrac{1}{2} c_{44}[(u_{2,3}^2 + u_{2,3} u_{3,2} + u_{3,2} u_{2,3} + u_{3,2}^2) + \cdots],\end{aligned}\right\} \tag{11.6}$$

we have

$$\left.\begin{aligned}c_{1111} = c_{2222} = c_{3333} = c_{11}, \\ c_{1122} = c_{2233} = c_{3311} = c_{2211} = c_{3322} = c_{1133} = c_{12}, \\ c_{pqpq} = c_{pqqp} = c_{qppq} = c_{qpqp} = c_{44} \quad \text{when} \quad q \neq p;\end{aligned}\right\} \tag{11.7}$$

and the other c_{ijkl}'s are zero.

For the G_{ij} term, we have (BECKER and DÖRING [*1*], p. 136)

$$\left.\begin{aligned}G_{ij}(\alpha_m) u_{i,j} = \varrho_0 g_{ij}(\alpha_m) u_{i,j} = k_0(e_1 + e_2 + e_3) + \\ + k_1(e_1 \alpha_1^2 + e_2 \alpha_2^2 + e_3 \alpha_3^2) + 2k_2(e_4 \alpha_2 \alpha_3 + e_5 \alpha_3 \alpha_1 + e_6 \alpha_1 \alpha_2) + \\ + k_3(e_1 + e_2 + e_3)(\alpha_1^4 + \alpha_2^4 + \alpha_3^4) + k_4(e_1 \alpha_1^4 + e_2 \alpha_2^4 + e_3 \alpha_3^4) + \\ + k_6(e_4 \alpha_1^2 \alpha_2 \alpha_3 + e_5 \alpha_2^2 \alpha_3 \alpha_1 + e_6 \alpha_3^2 \alpha_1 \alpha_2),\end{aligned}\right\} \tag{11.8}$$

if we neglect terms of higher than the fourth degree in the α's. On replacing e_1 by $u_{1,1}$, e_4 by $u_{2,3}+u_{3,2}$, etc. we find

$$\left.\begin{aligned} G_{11}(\alpha_m) &= k_0 + k_1\alpha_1^2 + k_3(\alpha_1^4+\alpha_2^4+\alpha_3^4) + k_4\alpha_1^4, \ldots, \\ G_{23}(\alpha_m) &= G_{32}(\alpha_m) = 2k_2\alpha_2\alpha_3 + k_6\alpha_1^2\alpha_2\alpha_3, \ldots, \end{aligned}\right\} \tag{11.9}$$

where in each case two other equations follow by cyclic permutation. Often it is sufficient to omit the fourth-degree terms, so that

$$G_{11}(\alpha_m) = k_0 + k_1\alpha_1^2, \ldots, \qquad G_{23}(\alpha_m) = G_{32}(\alpha_m) = 2k_2\alpha_2\alpha_3, \ldots. \tag{11.10}$$

The value of k_0 depends on the definition of zero strain; often this is so defined that

$$k_0 = -\tfrac{1}{3}k_1. \tag{11.11}$$

Then

$$G_{11}(\alpha_m) = k_1(\alpha_1^2 - \tfrac{1}{3}), \ldots. \tag{11.12}$$

The anisotropy-energy density of the undeformed material is (BECKER and DÖRING [*1*], p. 114; KNELLER [*1*], p. 180)

$$G(\alpha_m) = \varrho_0\, g(\mathsf{M}_P) = K_1(\alpha_1^2\alpha_2^2 + \alpha_2^2\alpha_3^2 + \alpha_3^2\alpha_1^2) + K_2\alpha_1^2\alpha_2^2\alpha_3^2, \tag{11.13}$$

so that

$$\frac{\partial G(\alpha_m)}{\partial \alpha_1} = 2\alpha_1[K_1(\alpha_2^2+\alpha_3^2) + K_2\alpha_2^2\alpha_3^2], \ldots. \tag{11.14}$$

The exchange terms are

$$\left.\begin{aligned} \varrho_0\,\mathsf{F}_{ex} &= \tfrac{1}{2}\{(b_{ij} + b_{klij}u_{(k,l)})\alpha_{m,i}\alpha_{m,j}\} \\ &= \tfrac{1}{2}[C + C'(e_1+e_2+e_3)]\alpha_{m,i}\alpha_{m,i} \\ &= \tfrac{1}{2}[C + C'(e_1+e_2+e_3)]\,[(\nabla\alpha_1)^2 + (\nabla\alpha_2)^2 + (\nabla\alpha_3)^2]. \end{aligned}\right\} \tag{11.15}$$

Thus

$$b_{ij} = C\,\delta_{ij}, \qquad b_{klij} = C'\,\delta_{ij}\,\delta_{kl}. \tag{11.16}$$

The formula for t_{ij}, Eq. (9.11), becomes for a cubic crystal

$$\left.\begin{aligned} t_{11} = {} & \tfrac{1}{2}C'[(\nabla\alpha_1)^2 + (\nabla\alpha_2)^2 + (\nabla\alpha_3)^2] + \\ & + [k_0 + k_1\alpha_1^2 + k_3(\alpha_1^4+\alpha_2^4+\alpha_3^4) + k_4\alpha_1^4] + \\ & + c_{11}e_1 + c_{12}(e_2+e_3), \ldots, \end{aligned}\right\} \tag{11.17}$$

$$\left.\begin{aligned} \left.\begin{matrix} t_{23} \\ t_{32} \end{matrix}\right\} = {} & \pm(K_1 + K_2\alpha_1^2)\alpha_2\alpha_3(\alpha_2^2 - \alpha_3^2) + \\ & + [k_2\alpha_2\alpha_3 + \tfrac{1}{2}k_6\alpha_1^2\alpha_2\alpha_3] + c_{44}e_4, \ldots. \end{aligned}\right\} \tag{11.18}$$

To apply these formulas rigorously, it would be necessary first to solve the zeroth-order (in $\boldsymbol{u}$) magnetic equilibrium equations, to find α_1, α_2, and α_3 as functions of position. These values must then be inserted in the above formulas for t_{ij}, and the partial differential equations $t_{ij,j} + M_s \alpha_j H_{i,j} = 0$ [Eq. (9.33)] must be solved for u_1, u_2, and u_3; the integration constants in the solution must be adjusted to satisfy the boundary conditions on u_i or on T_i [Eq. (9.34)]. After this it would in principle be possible to find a first-order (in $\boldsymbol{u}$) correction to the equilibrium magnetization distribution. But obviously the parts of the calculation previous to this step are so complicated that there is little chance of carrying them out in any but specially simple cases.

A more practical procedure is to constrain the α_i's and u_i's to some mathematical form, with undetermined coefficients; insert these expressions in the formulas for F and W_m; and minimize G approximately by minimization with respect to the coefficients. In this procedure, the exact expression for W_m may be replaced by approximate values based on maximization of W_{MH} or minimization of W_{MB} (see § 7.4), with $\boldsymbol{H}_1$ or $\boldsymbol{B}_1$ constrained to a convenient form.

11.3. Linear approximations. We now suppose that the magnetization $\boldsymbol{M}$ is so nearly along the X_3 (or z) axis that its transverse components, $M_s \alpha_1$ and $M_s \alpha_2$, may be treated as first-order small quantities: more explicity, we suppose that terms of higher than second order in α_1, α_2, u_1, u_2, u_3, and quantities derived from them by linear operations (*e.g.* differentiation and integration) may be neglected in the free energy G. The approximation now being made is in addition to, and presupposes, the small-displacement approximations of § 9.

The situation now being considered can be brought about by application of a large applied field along Oz. It can also exist, without the presence of such a field, if Oz is a direction of minimum anisotropy-energy density and if the specimen is of proper shape (elongated along Oz) and has been previously magnetized along Oz by application of a suitable field. For a cubic crystal, the present x_1, x_2, x_3 axes may differ from the axes (cube edges, [*100*] etc.) used in § 11.2. Thus if the present x_3 axis is to coincide with a direction of minimum anisotropy energy, and if the anisotropy energy density is $K_1(\alpha_1^2\alpha_2^2 + \alpha_2^2\alpha_3^2 + \alpha_3^2\alpha_1^2)$ in cubic axes (Eq. (11.13) with $K_2 = 0$), then for $K_1 > 0$ the new x_3 axis must coincide with one of the cubic x_i axes, but for $K_1 < 0$ it must coincide with one of the eight [*111*] directions $\pm\boldsymbol{i} \pm \boldsymbol{j} \pm \boldsymbol{k}$ in the cubic axes.

To the second order in α_1 and α_2,

$$\alpha_3 = [1 - (\alpha_1^2 + \alpha_2^2)]^{\frac{1}{2}} = 1 - \tfrac{1}{2}(\alpha_1^2 + \alpha_2^2); \qquad (11.19)$$

this formula may be used to eliminate α_3 wherever it occurs.

In § 9.3, we expressed G as $G^{(0)} + G^{(1)} + G^{(2)}$, where $G^{(n)}$ was of order n in u_i. In the approximation now to be made, $G^{(0)}$ must be expressed to the second order in α_1 and α_2, $G^{(1)}$ to the first, and $G^{(2)}$ to the zeroth.

The term $G^{(0)}$ is given by Eq. (4.22) with all quantities evaluated at $u_i = 0$. The expansion to the second order in α_1 and α_2 is the same as for a rigid body (BROWN [11], § 4.6; in the present application, we are generalizing the C term and supposing that the surface-anisotropy constant K_s vanishes). The result is (except for an additive constant)

$$\left.\begin{aligned} G^{(0)} = \int \Big\{ \frac{1}{2} b_{ij}\, \alpha_{p,i}\, \alpha_{p,j} + g_p\, \alpha_p + \frac{1}{2} g_{pq}\, \alpha_p\, \alpha_q + \\ + \frac{1}{2} M_s\, H_z^{(0)}\, \alpha_p\, \alpha_p - M_s\, H_p^{(0)}\, \alpha_p \Big\}\, d\tau_0 + \frac{1}{2\gamma} \int \boldsymbol{h}^2\, d\tau_0. \end{aligned}\right\} \quad (11.20)$$

Here i and j are summed over 1, 2, 3, but p and q are summed over 1 and 2 only; g_p and g_{pq} are defined by the statement that the anisotropy-energy density, to the second order in α_1 and α_2, is

$$\varrho_0\, g(\mathsf{M}_P) = \text{const} + g_p\, \alpha_p + \tfrac{1}{2} g_{pq}\, \alpha_p\, \alpha_q; \quad (11.21)$$

$H_1^{(0)}$, $H_2^{(0)}$, and $H_3^{(0)}$ $(= H_z^{(0)})$ are the components of magnetizing force in the reference state $\boldsymbol{M} = M_s \boldsymbol{k}$ (uniform magnetization along Oz); $\boldsymbol{h}$ is the magnetizing force due to the transverse magnetization $M_s(\alpha_1 \boldsymbol{i} + \alpha_2 \boldsymbol{j})$, and the $\boldsymbol{h}^2$ term in Eq. (11.20) is integrated thruout all space. We shall suppose that the specimen is an ellipsoid with principal axes along the coordinate axes; then

$$H_1^{(0)} = H_{01}, \quad H_2^{(0)} = H_{02}, \quad H_3^{(0)} = H_z^{(0)} = H_{03} - \gamma D_3 M_s, \quad (11.22)$$

where H_{0i} are the components of the applied field $\boldsymbol{H}_0$ (supposed uniform), and where D_3 is the demagnetizing factor (so defined that $D_1 + D_2 + D_3 = 1$ rather than 4π) for magnetization along Oz.

The term $G^{(1)}$ is given by Eq. (9.26); we shall suppose that $f_i = 0$ (no body forces). We require an expansion to the first order, only, in α_1 and α_2. By Eq. (11.19), $\alpha_{3,C}$ is of second order and $\alpha_{3,C}\, \alpha_{3,D}$ of fourth order; since $\alpha_{p,C}\, \alpha_{p,D}$ $(p = 1$ or $2)$ is of second order, the whole term containing $\mathsf{M}_{i,C}\, \mathsf{M}_{i,D}$ drops out. The term derived from $g(\mathsf{M}_P)$ reduces to the volume integral of $g_q(\alpha_p\, u_{[p,q]} + u_{[3,q]}) + g_{pq}\, \alpha_q\, u_{[3,p]}$, where g_q and g_{pq} have been defined by Eq. (11.21). The term derived from $g_{AB}(\mathsf{M}_P)$ reduces to the volume integral of $[q_{AB0} + q_{ABp}\, \alpha_p]\, u_{(A,B)}$, where q_{AB0} and q_{ABp} are defined by the statement that to the first order in α_1 and α_2,

$$\varrho_0\, g_{AB}(M_P) = q_{AB0} + q_{ABp}\, \alpha_p. \quad (11.23)$$

To evaluate $M_j\, H_{i,j}$ to the first order in α_1 and α_2, we first note that the zero-order part of $\boldsymbol{H}$, given by Eqs. (11.22), is uniform and therefore contributes nothing. Then since the zero-order part of $\boldsymbol{M}$ is $M_s \boldsymbol{k}$, $M_j\, H_{i,j}$ reduces to $M_s\, \partial h_i / \partial z$, where $\boldsymbol{h}$ is again, as in Eq. (11.20), the magnetizing force due to the first-order magnetization $M_s(\boldsymbol{i}\alpha_1+\boldsymbol{j}\,\alpha_2)$. The expansion of M_n^2 is straightforward. Thus

$$\left.\begin{aligned} G^{(1)} = \int\Big\{ & g_q(\alpha_p\, u_{[p,q]} + u_{[3,q]}) + g_{pq}\,\alpha_q\, u_{[3,p]} + \\ & + (q_{AB0} + q_{ABp}\,\alpha_p)\, u_{(A,B)} - M_s \frac{\partial \boldsymbol{h}}{\partial z}\cdot \boldsymbol{u}\Big\}\, d\tau_0 + \\ & + \int\Big\{-\frac{1}{2}\gamma\, M_s^2 (n_3^2 + 2 n_3\, n_p\, \alpha_p)\, n_i\, u_i - T_i\, u_i\Big\}\, d S_0 . \end{aligned}\right\} \quad (11.24)$$

The term $G^{(2)}$ is given by Eq. (9.27), in which $\varrho_0\, g_{ABCD}(\mathsf{M}_P)$ may be replaced by its value c_{ABCD} when $\alpha_1 = \alpha_2 = 0$ and $\alpha_3 = 1$.

$$G^{(2)} = \tfrac{1}{2}\int c_{ijkl}\, u_{(i,j)}\, u_{(k,l)}\, d\tau_0 . \quad (11.25)$$

In $G = G^{(0)} + G^{(1)} + G^{(2)}$, the superscript n in $G^{(n)}$ denotes the order of smallness in u_1, u_2, u_3, and quantities derived from them by linear operations. It is convenient at this point to perform a new separation of G,

$$G = \mathscr{G}^{(0)} + \mathscr{G}^{(1)} + \mathscr{G}^{(2)}, \quad (11.26)$$

into parts $\mathscr{G}^{(0)}$, $\mathscr{G}^{(1)}$, and $\mathscr{G}^{(2)}$ of orders 0, 1, and 2 respectively in the set of variables $\alpha_1, \alpha_2, u_1, u_2, u_3$, and quantities derived from them by linear operations. Then $\mathscr{G}^{(0)}$ is a constant in the variations to be performed and may therefore be disregarded;

$$\left.\begin{aligned} \mathscr{G}^{(1)} = & \int\{g_p\,\alpha_p - M_s\, H_p^{(0)}\,\alpha_p + g_q\, u_{[3,q]} + q_{ij0}\, u_{(i,j)}\}\, d\tau_0 + \\ & + \int\{-\tfrac{1}{2}\gamma\, M_s^2\, n_3^2\, n_i\, u_i - T_i\, u_i\}\, d S_0 ; \end{aligned}\right\} \quad (11.27)$$

and

$$\left.\begin{aligned} \mathscr{G}^{(2)} = \int\Big\{ & \frac{1}{2} b_{ij}\,\alpha_{p,i}\,\alpha_{p,j} + \frac{1}{2} g_{pq}\,\alpha_p\,\alpha_q + \frac{1}{2} M_s\, H_z^{(0)}\,\alpha_p\,\alpha_p + \\ & + g_q\,\alpha_p\, u_{[p,q]} + g_{pq}\,\alpha_q\, u_{[3,q]} + q_{ijp}\,\alpha_p\, u_{(i,j)} - \\ & - M_s(\partial h_i/\partial z)\, u_i + \frac{1}{2} c_{ijkl}\, u_{(i,j)}\, u_{(k,l)}\Big\}\, d\tau_0 + \\ & + \frac{1}{2\gamma}\int \boldsymbol{h}^2\, d\tau_0 + \int(-\gamma\, M_s^2)\, n_3\, n_p\,\alpha_p\, n_i\, u_i\, d S_0 . \end{aligned}\right\} \quad (11.28)$$

To find the magnetic equilibrium conditions, we vary α_1 and α_2 and set $\delta\mathscr{G}^{(1)} + \delta\mathscr{G}^{(2)} = 0$; the variations of α_1 and α_2 are arbitrary and independent. The variations of $\mathscr{G}^{(1)}$ and $\mathscr{G}^{(2)}$ are

$$\delta\mathscr{G}^{(1)} = \int\{g_p - M_s\, H_p^{(0)}\}\,\delta\alpha_p\, d\tau_0 , \quad (11.29)$$

$$\left.\begin{aligned} \delta\mathscr{G}^{(2)} = \int\{ & b_{ij}\,\alpha_{p,i}\,\delta\alpha_{p,j} + g_{pq}\,\alpha_q\,\delta\alpha_p + \\ & + g_r\, u_{[p,r]}\,\delta\alpha_p + g_{rp}\, u_{[3,r]}\,\delta\alpha_p + q_{ijp}\, u_{(i,j)}\,\delta\alpha_p + \\ & + M_s\, H_z^{(0)}\,\alpha_p\,\delta\alpha_p - M_s\,\boldsymbol{u}\cdot\partial(\delta\boldsymbol{h})/\partial z - M_s\, h_p\,\delta\alpha_p\}\, d\tau_0 - \\ & - \int\gamma\, M_s^2\, n_3\, n_p\, n_i\, u_i\,\delta\alpha_p\, d S_0 . \end{aligned}\right\} \quad (11.30)$$

The first and seventh integrals in $\delta\mathscr{G}^{(2)}$ must be transformed so as to replace the variations of $\alpha_{p,j}$ and $\partial\boldsymbol{h}/\partial z$ by the variations of the independent variables α_p. The first of these integrals can be transformed in the usual manner; if b_{ij} is independent of position within the specimen,

$$\left.\begin{aligned}\int b_{ij}\,\alpha_{p,i}\,\delta\alpha_{p,j}\,d\tau_0 = -\int b_{ij}\,\alpha_{p,ij}\,\delta\alpha_p\,d\tau_0 + \\ +\int b_{ij}\,\alpha_{p,i}\,n_j\,\delta\alpha_p\,d S_0.\end{aligned}\right\} \quad (11.31)$$

Transformation of the integral $-\int M_s\,\boldsymbol{u}\cdot\partial(\delta\boldsymbol{h}/\partial z)\,d\tau_0$ presents more of a problem. For this purpose we need the following magnetic reciprocity theorem, proved in Appendix B: if magnetization distribution $\boldsymbol{M}_1(x, y, z)$ produces magnetizing force (pole field) $\boldsymbol{H}_1(x, y, z)$ and if magnetization distribution $\boldsymbol{M}_2(x, y, z)$ produces magnetizing force $\boldsymbol{H}_2(x, y, z)$, then

$$\int\left[\boldsymbol{M}_1\cdot\frac{\partial\boldsymbol{H}_2}{\partial z}+\boldsymbol{M}_2\cdot\frac{\partial\boldsymbol{H}_1}{\partial z}\right]d\tau_0 = -\gamma\int M_{1\mathrm{n}}\,M_{2\mathrm{n}}\,n_3\,d S_0. \quad (11.32)$$

To apply this theorem, we set

$$\boldsymbol{M}_1 = M_s\,\delta\boldsymbol{v}, \qquad \boldsymbol{M}_2 = \boldsymbol{u}, \quad (11.33)$$

where

$$\boldsymbol{v} \equiv \alpha_1\,\boldsymbol{i} + \alpha_2\,\boldsymbol{j}. \quad (11.34)$$

Then

$$\boldsymbol{H}_1 = \delta\boldsymbol{h}, \qquad \boldsymbol{H}_2 = \boldsymbol{\eta}, \quad (11.35)$$

where $\boldsymbol{\eta}$ is the magnetizing force produced by magnetization $\boldsymbol{u}$; that is, $\boldsymbol{\eta}$ is the field intensity calculated from volume and surface pole densities $-\nabla\cdot\boldsymbol{u}$ and $\boldsymbol{n}\cdot\boldsymbol{u}$ respectively. (Those who dislike setting a magnetization equal to an elastic displacement may either insert an appropriate dimensional factor or reword the definition of $\boldsymbol{\eta}$ so as to make it purely mathematical.) The theorem then gives

$$\int\left[M_s\,\delta\boldsymbol{v}\cdot\frac{\partial\boldsymbol{\eta}}{\partial z}+\boldsymbol{u}\cdot\frac{\partial\delta\boldsymbol{h}}{\partial z}\right]d\tau_0 = -\gamma\int M_s\,u_{\mathrm{n}}\,\delta v_{\mathrm{n}}\,n_3\,d S_0, \quad (11.36)$$

or

$$\left.\begin{aligned}-\int M_s\,\boldsymbol{u}\cdot\frac{\partial(\delta\boldsymbol{h})}{\partial z}\,d\tau_0 &= \int M_s^2\,\frac{\partial\boldsymbol{\eta}}{\partial z}\cdot\delta\boldsymbol{v}\,d\tau_0 + \gamma\int M_s^2\,u_{\mathrm{n}}\,n_3\,\delta v_{\mathrm{n}}\,d S_0 \\ &= \int M_s^2\,\frac{\partial\eta_p}{\partial z}\,\delta\alpha_p\,d\tau_0 + \gamma\int M_s^2\,n_i\,u_i\,n_3\,n_p\,\delta\alpha_p\,d S_0.\end{aligned}\right\} \quad (11.37)$$

It will be noticed that the surface term in this expression cancels the last term in Eq. (11.30).

By introducing the transformations (11.31) and (11.37) into Eq. (11.30), we obtain an expression for $\delta\mathscr{G}^{(2)}$ in which the only variations appearing are those of α_1 and α_2.

On setting the coefficient of $\delta\alpha_p$ equal to zero in $\delta\mathscr{G}^{(1)}+\delta\mathscr{G}^{(2)}$, we get the magnetic equilibrium conditions

$$\left.\begin{aligned}\{g_p - M_s H_p^{(0)}\} + \Big\{- b_{ij}\,\alpha_{p,ij} + g_{pq}\,\alpha_q + g_r\,u_{[p,r]} + g_{rp}\,u_{[3,r]} + \\ + q_{ijp} u_{(i,j)} + M_s H_z^{(0)}\,\alpha_p - M_s h_p + M_s^2 \frac{\partial \eta_p}{\partial z}\Big\} = 0 \quad \text{in } V,\end{aligned}\right\} \tag{11.38}$$

$$b_{ij}\,\alpha_{p,i}\,n_j = 0 \quad \text{on } S. \tag{11.39}$$

These must be supplemented by the equations that determine $\boldsymbol{h}$ and $\boldsymbol{\eta}$. We may set

$$\boldsymbol{h} = -\nabla\varphi, \qquad \boldsymbol{\eta} = -\nabla\psi; \tag{11.40}$$

then the equations that determine φ and ψ may be written

$$\nabla^2\varphi = \gamma M_s \nabla\cdot\boldsymbol{v} = \gamma M_s\,\alpha_{p,p}, \qquad \nabla^2\psi = \gamma\nabla\cdot\boldsymbol{u} = \gamma\,u_{i,i} \quad \text{in } V, \tag{11.41}$$

$$\nabla^2\varphi = \nabla^2\psi = 0 \quad \text{outside } V, \tag{11.42}$$

$$\left.\begin{aligned}\left(-\frac{\partial\varphi}{\partial n} + \gamma M_s n_p\,\alpha_p\right)_{\text{in}} &= \left(-\frac{\partial\varphi}{\partial n}\right)_{\text{out}}, \\ \left(-\frac{\partial\psi}{\partial n} + \gamma\,n_i\,u_i\right)_{\text{in}} &= \left(-\frac{\partial\psi}{\partial n}\right)_{\text{out}} \quad \text{on } S,\end{aligned}\right\} \tag{11.43}$$

$$\varphi \text{ and } \psi \text{ regular at infinity.} \tag{11.44}$$

Alternatively, φ and ψ (or $\boldsymbol{h}$ and $\boldsymbol{\eta}$) may be expressed directly as integrals over their sources.

To find the mechanical equilibrium equations, we vary u_1, u_2, and u_3 and set $\delta\mathscr{G}^{(1)}+\delta\mathscr{G}^{(2)}=0$; the variations of the u_i's are arbitrary and independent. In this case the transformations necessary to eliminate variations $\delta u_{i,j}$, leaving only variations δu_i, are standard ones. We suppose that g_r, q_{ij0}, etc. are independent of position within the specimen; then the transformations just mentioned make $\mathscr{G}^{(1)}$ entirely a surface integral. We need not write the long and complicated expression for $\delta\mathscr{G}^{(1)}+\delta\mathscr{G}^{(2)}$, since it can easily be reconstructed from the equilibrium equations. They are: in V,

$$\left.\begin{aligned}&-\frac{1}{2} g_r\,\alpha_{1,r} + \frac{1}{2}\alpha_{p,p}\,g_1 + \frac{1}{2} g_{1q}\,\alpha_{q,3} - \\ &\qquad - \left[q_{(1j)p}\,\alpha_{p,j} + c_{1jkl}\,u_{k,lj} + M_s\frac{\partial h_1}{\partial z}\right] = 0, \\ &-\frac{1}{2} g_r\,\alpha_{2,r} + \frac{1}{2}\alpha_{p,p}\,g_2 + \frac{1}{2} g_{2q}\,\alpha_{q,3} - \\ &\qquad - \left[q_{(2j)p}\,\alpha_{p,j} + c_{2jkl}\,u_{k,lj} + M_s\frac{\partial h_2}{\partial z}\right] = 0, \\ &-\frac{1}{2} g_{rq}\,\alpha_{q,r} - \left[q_{(3j)p}\,\alpha_{p,j} + c_{3jkl}\,u_{k,lj} + M_s\frac{\partial h_3}{\partial z}\right] = 0;\end{aligned}\right\} \tag{11.45}$$

on S,

$$\left.\begin{aligned}
&-\tfrac{1}{2}n_3 g_1+[q_{(1j)0}\,n_j-\tfrac{1}{2}\gamma M_s^2 n_3^2 n_1-T_1]+\\
&\quad+\{\tfrac{1}{2}g_r n_r \alpha_1-\tfrac{1}{2}\alpha_p n_p g_1-\tfrac{1}{2}n_3 g_{1q}\alpha_q+\\
&\quad+[q_{(1j)p}\,\alpha_p n_j+c_{1jkl}\,u_{k,l}\,n_j-\gamma M_s^2 n_3 n_p \alpha_p n_1]\}=0,\\
&-\tfrac{1}{2}n_3 g_2+[q_{(2j)0}\,n_j-\tfrac{1}{2}\gamma M_s^2 n_3^2 n_2-T_2]+\\
&\quad+\{\tfrac{1}{2}g_r n_r \alpha_2-\tfrac{1}{2}\alpha_p n_p g_2-\tfrac{1}{2}n_3 g_{2q}\alpha_q+\\
&\quad+[q_{(2j)p}\,\alpha_p n_j+c_{2jkl}\,u_{k,l}\,n_j-\gamma M_s^2 n_3 n_p \alpha_p n_2]\}=0,\\
&\tfrac{1}{2}(g_1 n_1+g_2 n_2)+[q_{(3j)0}\,n_j-\tfrac{1}{2}\gamma M_s^2 n_3^3-T_3]+\\
&\quad+\{\tfrac{1}{2}g_{rq}\alpha_q n_r+[q_{(3j)p}\,\alpha_p n_j+c_{3jkl}\,u_{k,l}\,n_j-\gamma M_s^2 n_3^2 n_p \alpha_p]\}=0.
\end{aligned}\right\}\qquad(11.46)$$

Summations of $i, j, k, \ldots$ are over 1, 2, 3; of $p, q, r, \ldots$, over 1 and 2 only. The usual permissible symmetries of coefficients (e.g. $c_{ijkl}=c_{jikl}=c_{klij}$) have been assumed.

For a rigid body, these linearized equations occur in two types of problem. In the first type, a large field $\boldsymbol{H}_0=H_0\boldsymbol{k}$ is present, but complete saturation is prevented by random transverse forces, and the resulting deviations from saturation, in the equilibrium state at given $\boldsymbol{H}_0=H_0\boldsymbol{k}$, are to be studied. Here $H_p^{(0)}$ in Eq. (11.38) is zero, but g_p is not. The approach to saturation has been studied by this method (Brown [*3*], [*4*]; Seeger and Kronmüller [*1*]). In particular, the effect of dislocations on magnetization was investigated, but by an approximate method: the stress field of the dislocations was calculated as for a non-magnetic body, the stresses thus found were inserted into formulas of the conventional magnetostriction theory to find effective transverse forces $-g_p$ that might be considered to act on the local magnetization vector, and the magnetization distribution [i.e. the functions $\alpha_p(x, y, z)$] was found as for a rigid body. The present theory, in principle, permits a self-consistent calculation, in which magnetoelastic interactions are taken into account at every stage — including the calculation of the deformation field (and consequent stress field, however defined) about a dislocation. The calculation will evidently be extremely complicated.

In the second type of problem, the conditions are such that the state $\alpha_1=\alpha_2=0$ is an equilibrium state, and the stability of this state is to be studied. When $u_i=0$ (*i.e.* for a rigid body), the condition for equilibrium at $\alpha_1=\alpha_2=0$ is the vanishing of $g_p-M_s H_p^{(0)}$ in Eq. (11.38); and in the simpler cases (an ellipsoid with a principal axis, a direction of minimum anisotropy energy, and the applied field $\boldsymbol{H}_0$ all along Oz), the two terms g_p and $M_s H_p^{(0)}$ vanish separately. The equilibrium is obviously stable at sufficiently large positive values of H_0 and unstable at sufficiently large negative; for then the third term in Eq. (11.28),

$\frac{1}{2} M_s H_z^{(0)} \int (\alpha_1^2 + \alpha_2^2)\, d\tau_0$, dominates and determines the sign of G. The problem is to find at what value of H_0, as it is decreased from large positive to large negative values, the equilibrium becomes unstable. Altho the most convincing formulation of this problem is one based on the sign of $\delta^2 G$, an equivalent and simpler one is the following: at the point where the equilibrium changes from stable to unstable, it is neutral; and therefore at this particular H_0, some state with α_1 or α_2 (or both) $\neq 0$, as well as the state $\alpha_1 = \alpha_2 = 0$, must also be an equilibrium state. The problem is therefore to determine the algebraically largest H_0 at which the *homogeneous* system of equations (11.38) to (11.44) with $g_p - M_s H_p^{(0)} = 0$ (and with $u_i = 0$) possesses a nontrivial solution. This value of H_0 is known as the *nucleation field*. Besides its value, the corresponding functions $\alpha_p(x, y, z)$ are of interest; they are of course, indeterminate by an arbitrary constant factor, since the equations that determine them are linear and homogeneous.

With deformations allowed, the problem becomes considerably more complicated. We shall examine some specially simple cases of it in § 11.4.

11.4. The nucleation-field problem. We shall consider the nucleation-field problem for the special case of a homogeneous cubic crystal in the form of an ellipsoid, with a principal ellipsoid axis and the applied field both along a [100] direction. We may then choose the x_1, x_2, x_3 axes of § 11.3 to coincide with those of § 11.2. Our first problem is to evaluate the constants in Eqs. (11.27) and (11.28) for this special case.

By the first Eq. (11.16),

$$b_{ij} = C\, \delta_{ij}; \tag{11.47}$$

we do not need b_{klij}.

By expanding Eq. (11.13) to the second order in α_1 and α_2, we find according to the defining equation (11.21)

$$\left.\begin{aligned} g_1 &= g_2 = 0, \\ g_{11} &= g_{22} = 2K_1, \\ g_{12} &= g_{21} = 0. \end{aligned}\right\} \tag{11.48}$$

By expanding Eq. (11.8) to the first order in α_1 and α_2, we find according to the defining equation (11.23)

$$\left.\begin{aligned} q_{110} &= q_{220} = k_0 + k_3, \\ q_{330} &= k_0 + k_1 + k_3 + k_4, \\ q_{232} &= q_{322} = q_{311} = q_{131} = 2k_2; \end{aligned}\right\} \tag{11.49}$$

the remaining q_{ij0}'s and q_{ijp}'s vanish.

Insertion of these values into Eqs. (11.27) and (11.28) gives (since $H_1^{(0)} = H_2^{(0)} = 0$)

$$\left.\begin{aligned}\mathcal{G}^{(1)} &= \int\Big\{(k_0 + k_3)\,(u_{(1,1)} + u_{(2,2)}) + (k_0 + k_1 + k_3 + k_4)\,u_{(3,3)}\Big\}\,d\tau_0 + \\ &+ \int\Big\{-\frac{1}{2}\gamma M_s^2\, n_3^2\, u_n - T_i\, u_i\Big\}\,dS_0,\end{aligned}\right\} \quad (11.50)$$

$$\left.\begin{aligned}\mathcal{G}^{(2)} &= \int\Big\{\frac{1}{2}C\,\alpha_{p,i}\,\alpha_{p,i} + K_1(\alpha_1^2 + \alpha_2^2) + \\ &+ 2K_1\,\alpha_1\,u_{[3,1]} + 2K_1\,\alpha_2\,u_{[3,2]} + \\ &+ 2k_2\,[\alpha_2\,u_{(2,3)} + \alpha_2\,u_{(3,2)} + \alpha_1\,u_{(3,1)} + \alpha_1\,u_{(1,3)}] + \\ &+ \frac{1}{2}C_{ijkl}\,u_{i,j}\,u_{k,l} + \frac{1}{2}M_s\,H_z^{(0)}(\alpha_1^2 + \alpha_2^2) - \\ &- M_s\,\frac{\partial \boldsymbol{h}}{\partial z}\cdot\boldsymbol{u}\Big\}\,d\tau_0 + \frac{1}{2\gamma}\int \boldsymbol{h}^2\,d\tau_0 - \\ &- \int \gamma M_s^2\, n_3(n_1\,\alpha_1 + n_2\,\alpha_2)\,u_n\,dS_0.\end{aligned}\right\} \quad (11.51)$$

The elastic constants (in the approximation described in § 11.2, which we retain here) are given by Eqs. (11.7), but there is no advantage in introducing their values explicitly at this point.

The equilibrium equations may be found either by variation of Eqs. (11.50) and (11.51), or by substitution of Eqs. (11.47) to (11.49) in the general equilibrium equations (11.38), (11.39), (11.45), and (11.46). The magnetic equilibrium equations are: in V,

$$\left.\begin{aligned}-C\,\nabla^2\alpha_p + [2K_1 + M_s\,H_z^{(0)}]\,\alpha_p - M_s\,h_p + 2K_1\,u_{[3,p]} + \\ + 4k_2\,u_{(p,3)} + M_s^2\,\partial\eta_p/\partial z = 0 \quad (p = 1, 2);\end{aligned}\right\} \quad (11.52)$$

on S,

$$C\,\partial\alpha_1/\partial n = C\,\partial\alpha_2/\partial n = 0. \quad (11.53)$$

The elastic equilibrium equations are: in V,

$$c_{pjkl}\,u_{k,lj} + (2k_2 - K_1)\,\alpha_{p,3} + M_s\,\partial h_p/\partial z = 0 \quad (p = 1, 2), \quad (11.54)$$

$$c_{3jkl}\,u_{k,lj} + (2k_2 + K_1)\,(\alpha_{1,1} + \alpha_{2,2}) + M_s\,\partial h_3/\partial z = 0; \quad (11.55)$$

on S,

$$\left.\begin{aligned}c_{pjkl}\,u_{k,l}\,n_j + (2k_2 - K_1)\,\alpha_p\,n_3 - \gamma M_s^2\,n_3\,n_p(n_1\,\alpha_1 + n_2\,\alpha_2) \\ = T_p + \tfrac{1}{2}\gamma M_s^2\,n_3^2\,n_p - (k_0 + k_3)\,n_p \quad (p = 1, 2),\end{aligned}\right\} \quad (11.56)$$

$$\left.\begin{aligned}c_{3jkl}\,u_{k,l}\,n_j + (2k_2 + K_1)\,(n_1\,\alpha_1 + n_2\,\alpha_2) - \gamma M_s^2\,n_3^2(n_1\,\alpha_1 + n_2\,\alpha_2) \\ = T_3 + \tfrac{1}{2}\gamma M_s^2\,n_3^3 - (k_0 + k_1 + k_3 + k_4)\,n_3.\end{aligned}\right\} \quad (11.57)$$

Eqs. (11.52) to (11.57), together with the auxiliary φ and ψ Eqs. (11.41) to (11.44), determine an equilibrium set of values of α_1, α_2, u_1, u_2, u_3 for given surface forces T_i, provided these T_i's satisfy the conditions stated in § 7.9. In the present case, with $\boldsymbol{H}_0$ uniform and with $f_i=0$, these reduce to the conditions that the total force $\int T_i\, dS_0$ must vanish and that the total couple must equilibrate the magnetic couple $\int \boldsymbol{M}\times\boldsymbol{H}_0\, d\tau_0$. In the present linear approximation, the second condition becomes

$$\left.\begin{aligned}\int (y\,T_3-z\,T_2)\,dS &= -H_0\,M_s\int \alpha_2\,d\tau,\\ \int (z\,T_1-x\,T_3)\,dS &= +H_0\,M_s\int \alpha_1\,d\tau,\\ \int (x\,T_2-y\,T_1)\,dS &= 0\end{aligned}\right\}\tag{11.58}$$

(we drop the subscript $_0$ on $d\tau$ and dS, since in the linear approximation the distinction between $d\tau_0$ and $d\tau$ is no longer important). To make the u_i's for given α_p's unique, we may require that at the origin (the center of the ellipsoid), $u_i=0$ and $u_{[i,j]}=0$.

We suppose that the body is mechanically free except for the surface forces necessary to satisfy the conditions (11.58). These conditions do not determine $\boldsymbol{T}$, since to any set of T_i's satisfying them we may add any set that gives zero total force and couple. As in many purely elastic problems, however, it is reasonable to suppose that any such additional set of T_i's will produce only local surface effects, not important in comparison with such volume averages as $\int \alpha_p\, d\tau$. We may then select T_i's that satisfy Eqs. (11.58) and are analytically simple. One such set of T_i's is

$$T_i=\sigma_{ij}\,n_j,\tag{11.59}$$

where σ_{ij} is a constant antisymmetric tensor. Eq. (11.59) gives

$$\int T_i\,dS=\sigma_{ij}\int n_j\,dS=\sigma_{ij}\int \partial 1/\partial x_j\,d\tau=0,\tag{11.60}$$

$$\left.\begin{aligned}\int x_k\,T_i\,dS &=\sigma_{ij}\int x_k\,n_j\,dS=\sigma_{ij}\int x_{k,j}\,d\tau\\ &=\sigma_{ij}\,\delta_{kj}\,V=\sigma_{ik}\,V.\end{aligned}\right\}\tag{11.61}$$

To satisfy Eqs. (11.58), we must take

$$\left.\begin{aligned}\sigma_{23}=-\sigma_{32}&=\frac{H_0\,M_s}{2V}\int \alpha_2\,d\tau,\\ \sigma_{13}=-\sigma_{31}&=\frac{H_0\,M_s}{2V}\int \alpha_1\,d\tau.\end{aligned}\right\}\tag{11.62}$$

Then

$$T_p=n_3\,\frac{H_0\,M_s}{2V}\int \alpha_p\,d\tau\quad (p=1,2),\tag{11.63}$$

$$T_3=-\frac{H_0\,M_s}{2V}\left\{n_1\int \alpha_1\,d\tau+n_2\int \alpha_2\,d\tau\right\}.\tag{11.64}$$

If these values of T_i are substituted in the mechanical boundary equations (11.56), (11.57) and transposed to the left, we get a set of partial differential equations and boundary conditions that would be homogeneous in the unknown functions α_1, α_2, u_1, u_2, u_3, φ, and ψ were it not for the remaining terms on the right in Eqs. (11.56), (11.57). These terms determine, for any H_0, an equilibrium solution that is the analog of the equilibrium solution $\alpha_1=\alpha_2=0$ in the rigid-body case. If we attempt a solution with $\alpha_1=\alpha_2=0$ but $u_i\neq 0$, the u_i's thereby determined will be those studied in §§ 10.2—10.4. Since these u_i's vary in a complicated manner with position (in a sphere, they are cubic functions of the coordinates), they will in general produce nonvanishing terms $4k_2\, u_{(p,3)}$ and $M_s^2\, \partial\eta_p/\partial z$ in the magnetic equilibrium equations (11.52); the assumed values $\alpha_1=\alpha_2=0$ will therefore not satisfy these equilibrium equations, and the attempt at such a solution fails. The actual equilibrium solution must involve a nonvanishing α_1 and α_2. (The u_i's calculated in §§ 10.2—10.4 are therefore strictly valid only in the limit as $H_0\to\infty$, when $\alpha_p\to 0$; they are good approximations, however, at large H_0.)

For determination of the nucleation field, the form of this equilibrium solution need not be known; what we are interested in is its stability with respect to small deviations. If $\alpha_p^{(0)}$, $u_i^{(0)}$, $\varphi^{(0)}$, $\psi^{(0)}$ are the equilibrium functions and $\alpha_p^{(1)}$, $u_i^{(1)}$, $\varphi^{(1)}$, $\psi^{(1)}$ are the deviations, neutrality of the equilibrium requires that $\alpha_p^{(0)}+\alpha_p^{(1)}$ etc. also satisfy the equilibrium equations; the deviations $\alpha_p^{(1)}$ etc. must therefore satisfy the corresponding *homogeneous* equations. This can occur only at certain special values of H_0; the algebraically largest such value is the nucleation field.

The nucleation field is therefore the algebraically largest eigenvalue of the parameter H_0 (recall that $H_z^{(0)}=H_0-\gamma D_3 M_s$) in the homogeneous system of linear equations consisting of Eqs. (11.52) to (11.55), Eqs. (11.41) to (11.44), and — instead of Eqs. (11.56), (11.57) — the homogeneous boundary conditions

$$\left.\begin{aligned} &c_{pjkl}\, u_{k,l}\, n_j + (2k_2 - K_1)\,\alpha_p\, n_3 - \gamma M_s^2\, n_3\, n_p (n_1\,\alpha_1 + n_2\,\alpha_2) - \\ &\qquad - n_3 \frac{H_0 M_s}{2V}\int \alpha_p\, d\tau = 0 \quad (p=1,2), \end{aligned}\right\} \tag{11.65}$$

$$\left.\begin{aligned} &c_{3jkl}\, u_{k,l}\, n_j + (2k_2 + K_1)\, (n_1\,\alpha_1 + n_2\,\alpha_2) - \gamma M_s^2\, n_3^2 (n_1\,\alpha_1 + n_2\,\alpha_2) + \\ &\qquad + \frac{H_0 M_s}{2V}\left\{n_1\int \alpha_1\, d\tau + n_2\int \alpha_2\, d\tau\right\} = 0; \end{aligned}\right\} \tag{11.66}$$

additional requirements are that u_i and $u_{[i,j]}$ vanish at $x_i=0$. The variables α_p, u_i, etc. that occur in this final formulation of the nucleation-field problem are, physically, the deviations $\alpha_p^{(1)}$, $u_i^{(1)}$, etc. from the reference equilibrium values $\alpha_p^{(0)}$, $u_i^{(0)}$, etc. that exist at any H_0.

The problem thus stated is a complicated one; it involves seven second-order partial differential equations (in α_1, α_2, u_1, u_2, u_3, φ, and ψ) in the region V inside S, two second-order partial differential equations (in φ and ψ) in the region outside S, nine boundary conditions on S, and two conditions of regularity at infinity. No attempt will be made here to solve this problem. We shall, however, make a few remarks about methods of simplifying the problem.

First, an approximate solution can be found by constraining the strains to be uniform and the rotations to be zero:

$$u_i = A_{ij} x_j, \qquad A_{ij} = A_{ji}. \tag{11.67}$$

If these constraints are introduced in $\mathcal{G}^{(1)}$ and $\mathcal{G}^{(2)}$, Eqs. (11.50), (11.51), the subsequent variational procedure will lead to a system of equations in three unknown functions α_1, α_2, and φ and in six unknown constants A_{ij}. Solution of this modified problem might prove considerably simpler than solution of the original problem. The approximate value obtained for $-H_0$ will be a lower bound to the correct value. If a method can be devised for finding an upper bound as well, the results may be sufficient for practical purposes.

Second, methods already used successfully for a rigid body (Brown [*9*], [*12*]; Aharoni [*1*]) can be applied to the functions α_1, α_2, and φ, again with the result that upper and lower bounds to $-H_0$ can be found.

Third, a somewhat less difficult problem occurs if, instead of specifying that the surface must be mechanically free except for the couple necessary to equilibrate the magnetic couple, we specify a fixed surface: $u_i = 0$ on S. If we review the thermodynamic arguments of § 7 with this case in mind, we find that the thermodynamic potential to be minimized differs from the previous one by omission of the terms containing f_i (which we here suppose to vanish) and T_i. Because of the boundary constraint $\delta u_i = 0$, the surface integrals that led to the previous surface equilibrium equations (11.56), (11.57) now vanish; instead of these boundary conditions, we now have the boundary conditions $u_i = 0$.

We shall examine, with these new boundary conditions, the effect of magnetostriction on the nucleation field for "rotation in unison": $\alpha_1 = \text{const}$, $\alpha_2 = \text{const}$. If we tentatively assume α_p's of this form, we have $h_1 = -\gamma M_s D_1 \alpha_1$, $h_2 = -\gamma M_s D_2 \alpha_2$, where the D_i's are the demagnetizing factors. Since both α_p and h_i are independent of position within the specimen, the mechanical equilibrium equations (11.54), (11.55) reduce to

$$c_{ijkl} u_{k,lj} = 0, \tag{11.68}$$

which are the same as for a nonmagnetic body with elastic constants c_{ijkl} under no body forces. With the boundary conditions $u_i = 0$ on S, the unique solution is $u_i = 0$ thruout V. The magnetic equilibrium equations (11.52), (11.53) then contain only the terms that occur in the corresponding rigid-body problem (since $u_{[3,p]} = u_{(p,3)} = 0$ and

$\eta_p = 0$). Therefore with the boundary conditions $u_i = 0$, magnetostriction has no effect on the nucleation field for rotation in unison.

The same conclusion follows for "curling" in an infinitely long circular cylinder. In this case $\alpha_1 = -v_\varphi(\varrho) \sin\varphi$, $\alpha_2 = v_\varphi(\varrho)\cos\varphi$, where ϱ, φ, z are cylindrical coordinates: there is only a circumferential component $M_s v_\varphi$ of the transverse magnetization, and v_φ depends only on distance ϱ from the cylinder axis. Such a transverse magnetization produces no volume or surface poles, so that $\boldsymbol{h} = 0$; also $\alpha_{p,3} = 0$ and $\alpha_{1,1} + \alpha_{2,2} = \nabla \cdot \boldsymbol{v} = 0$. Therefore Eqs. (11.54), (11.55) again reduce to Eq. (11.68) and, with the boundary condition $u_i = 0$ on S, determine that $u_i = 0$ thruout V. Eqs. (11.52), (11.53) then reduce to the form they had in the rigid-body curling calculation, and the nucleation field is unaffected.

For curling in a *sphere*, v_φ is a function of r and ϑ (in spherical coordinates r, ϑ, φ), so that $\alpha_{p,3}$ in Eq. (11.54) no longer vanishes [altho $\alpha_{1,1} + \alpha_{2,2} = \nabla \cdot \boldsymbol{v}$ in Eq. (11.55) does vanish]. In this case u_i will differ from zero at internal points, and the nucleation field will apparently be affected by magnetostriction.

The case $u_i = 0$ on S is unrealistic for an individual laboratory specimen; for fine particles in a solid nonmagnetic matrix, however, it is probably no worse an approximation than the free-surface case. A still better approximation would be to solve the elastic problem for the surrounding nonmagnetic material (with appropriate conditions at infinity) and to require continuity of u_i and τ_i across S.

At this point we shall leave the nucleation-field problem. Its solution in particular cases is clearly not a simple matter. In this monograph we have only laid the basis for a self-consistent attack on the nucleation-field problem and on other problems in which magnetoelastic interactions play a role. The actual solution of such problems remains for doctoral dissertations of the future.

Appendix A

The Variation of the Magnetic Self-Energy

In the following we consider variations $\delta \mathsf{M}_i$ of the moment per unit mass and variations δx_i of the coordinates. The variations may be either actual or virtual; in the former case, we get the time rate of change of W_m by replacing δ by (d/dt), or $\delta \mathsf{M}_i$, δx_i, and δW_m by $\dot{\mathsf{M}}_i$, $\dot{x}_i$, and $\dot{W}_m$ respectively.

For any function f, the notation $f_{,i}$ will imply that f is expressed as a function of the variables x_i before differentiation; the notation $f_{,A}$, that it is expressed as a function of the variables X_A. Thus if f was originally expressed as

$$f(x_1, x_2, x_3, X_1, X_2, X_3),$$

$f_{,i}$ means

$$\partial f\big(x_1, x_2, x_3, X_1(x_1, x_2, x_3),\ X_2(x_1, x_2, x_3),\ X_3(x_1, x_2, x_3)\big)/\partial x_i,$$

and $f_{,A}$ means

$$\partial f\big(x_1(X_1, X_2, X_3),\ x_2(X_1, X_2, X_3),\ x_3(X_1, X_2, X_3),\ X_1, X_2, X_3\big)/\partial X_A.$$

(TRUESDELL and TOUPIN [1] would write $f_{,i}$ for

$$\partial f(x_1, x_2, x_3, X_1, X_2, X_3)/\partial x_i,$$

$f_{,A}$ for

$$\partial f(x_1, x_2, x_3, X_1, X_2, X_3)/\partial X_A,$$

$f_{;i} = f_{,i} + f_{,A} X_{A,i}$ for our $f_{,i}$, and $f_{;A} = f_{,A} + f_{,i} x_{i,A}$ for our $f_{,A}$.)

The following auxiliary formulas will be needed:

$$n_i = J N_A X_{A,i}\, dS_0/dS, \tag{A.1}$$

$$(J X_{A,i})_{,A} = 0, \tag{A.2}$$

$$(\delta x_i)_{,A} = \delta(x_{i,A}) \equiv \delta x_{i,A} \qquad \text{etc.}, \tag{A.3}$$

$$\delta X_{A,j} \equiv \delta(X_{A,j}) = -X_{A,k} X_{C,j}\, \delta x_{k,C}, \tag{A.4}$$

$$\delta J = J X_{C,k}\, \delta x_{k,C}. \tag{A.5}$$

Eq. (A.1) is another form of Eq. (3.26); Eq. (A.2) is Eq. (7.26) with the roles of the variables x_i and X_A interchanged. Eq. (A.3) and its analog for M_i follow from the definitions of the variations, $\delta x_i(X_A) \equiv x_i'(X_A) - x_i(X_A)$ (where $x_i'(X_A)$ is the varied function), and enable us to dispense with the parentheses. To prove Eq. (A.4), we start with $X_{A,k}\, x_{k,C} = \delta_{AC}$ and get

$$x_{k,C}\, \delta X_{A,k} = -X_{A,k}\, \delta x_{k,C}. \tag{A.6}$$

Operation with $X_{C,j}$ gives, since $X_{C,j}\, x_{k,C} = \delta_{jk}$,

$$\delta_{jk}\, \delta X_{A,k} = -X_{C,j} X_{A,k}\, \delta x_{k,C}; \tag{A.7}$$

this reduces directly to Eq. (A.4).

To prove Eq. (A.5), we find from

$$J = \begin{vmatrix} x_{1,1} & x_{1,2} & x_{1,3} \\ x_{2,1} & x_{2,2} & x_{2,3} \\ x_{3,1} & x_{3,2} & x_{3,3} \end{vmatrix} \tag{A.8}$$

that

$$\delta J = K_{kC}\, \delta x_{k,C}, \tag{A.9}$$

where K_{kC} is the cofactor of $x_{k,C}$ in the determinant (A.8). If, for $i=1, 2, 3$ and for fixed k, we regard the three equations $x_{i,A} X_{A,k}=\delta_{ik}$ as determining the three unknowns $X_{1,k}$, $X_{2,k}$, and $X_{3,k}$ and solve for these unknowns, we get

$$X_{C,k}=K_{kC}/J. \tag{A.10}$$

On replacing K_{kC} in Eq. (A.9) by $J X_{C,k}$, we get Eq. (A.5).

We write $\boldsymbol{H}'$ (instead of $\boldsymbol{H}_1$) for the part of the magnetizing force due to the magnetization of the mass m under consideration, and we set $\boldsymbol{H}'=-\nabla\varphi$, or $H_i'=-\varphi_{,i}$. We start with W_{m} in the form

$$W_{\mathrm{m}}=\frac{1}{2\gamma}\int \boldsymbol{H}'^2\,d\tau=\frac{1}{2\gamma}\int \varphi_{,i}\,\varphi_{,i}\,d\tau. \tag{A.11}$$

The integral must be extended over all space. We suppose that the functions $x_i(X_A)$ are defined thruout the whole of space, in such a way that the values of x_i and $x_{i,A}$ outside the actual body are continuous with those inside (and vanish at infinity with sufficient rapidity to insure all necessary convergences). Then we may rewrite Eq. (A.11)

$$W_{\mathrm{m}}=\frac{1}{2\gamma}\int \varphi_{,A}\,\varphi_{,B}\,X_{A,i}\,X_{B,i}\,J\,d\tau_0. \tag{A.12}$$

On taking the variation, we get

$$\left.\begin{aligned}\delta W_{\mathrm{m}}&=\frac{1}{2\gamma}\int\{\varphi_{,A}\,\delta(\varphi_{,B}X_{A,i}X_{B,i}J)+(\varphi_{,B}X_{A,i}X_{B,i}J)\,\delta\varphi_{,A}\}\,d\tau_0\\&=\frac{1}{2\gamma}\int\{\varphi N_A\,\delta(\varphi_{,B}X_{A,i}X_{B,i}J)|_+^-+(\varphi_{,B}X_{A,i}X_{B,i}J)|_+^-N_A\,\delta\varphi\}\,dS_0-\\&\quad-\frac{1}{2\gamma}\int\{\varphi\,\delta(\varphi_{,B}X_{A,i}X_{B,i}J)_{,A}+(\varphi_{,B}X_{A,i}X_{B,i}J)_{,A}\,\delta\varphi\}\,d\tau_0,\end{aligned}\right\} \tag{A 13}$$

by the usual integration by parts. The notation $|_+^-$ means the internal value minus the external.

Now

$$\varphi_{,i}|_+^-\,n_i=\gamma M_i\,n_i. \tag{A.14}$$

With use of Eq. (A.1), this gives

$$\varphi_{,i}|_+^-N_A\,X_{A,i}=\gamma\varrho_0 J^{-1}\mathsf{M}_i\,N_A\,X_{A,i} \tag{A.15}$$

or

$$(\varphi_{,B}\,X_{A,i}\,X_{B,i}J)|_+^-N_A=\gamma\varrho_0\,\mathsf{M}_i\,N_A\,X_{A,i}. \tag{A.16}$$

Applying δ, we get

$$N_A\,\delta(\varphi_{,B}\,X_{A,i}\,X_{B,i}\,J)|_+^-=\gamma N_A\,\delta(\varrho_0\,\mathsf{M}_i\,X_{A,i}). \tag{A.17}$$

Therefore the surface terms in Eq. (A.13) become

$$\tfrac{1}{2}\int\{\varphi N_A\,\delta(\varrho_0\,\mathsf{M}_i\,X_{A,i})+\varrho_0\,\mathsf{M}_i\,N_A\,X_{A,i}\,\delta\varphi\}\,dS_0. \tag{A.18}$$

In the volume terms, we have

$$\varphi_{,ii} = \gamma M_{i,i}, \tag{A.19}$$

or

$$(\varphi_{,B} X_{B,i})_{,A} X_{A,i} = \gamma M_{i,A} X_{A,i} = \gamma (\varrho_0 J^{-1} \mathsf{M}_i)_{,A} X_{A,i}, \tag{A.20}$$

so that

$$(\varphi_{,B} X_{B,i})_{,A} X_{A,i} J = \gamma (\varrho_0 J^{-1} \mathsf{M}_i)_{,A} J X_{A,i}. \tag{A.21}$$

With use of Eq. (A.2), this becomes

$$(\varphi_{,B} X_{B,i} X_{A,i} J)_{,A} = \gamma (\varrho_0 \mathsf{M}_i X_{A,i})_{,A}. \tag{A.22}$$

Applying δ, we get

$$\delta (\varphi_{,B} X_{B,i} X_{A,i} J)_{,A} = \gamma \delta (\varrho_0 \mathsf{M}_i X_{A,i})_{,A}. \tag{A.23}$$

Therefore the volume terms in Eq. (A.13) become

$$-\tfrac{1}{2} \int \{\varphi \delta (\varrho_0 \mathsf{M}_i X_{A,i})_{,A} + (\varrho_0 \mathsf{M}_i X_{A,i})_{,A} \delta\varphi\} d\tau_0. \tag{A.24}$$

If the surface terms (A.18) and the volume terms (A.24) are added, and if the latter are then subjected to the usual transformation $\int f g_{,A} d\tau_0 = \int f g N_A dS_0 - \int f_{,A} g d\tau_0$, the new surface terms cancel the old ones, and we are left with

$$\delta W_{\mathrm{m}} = \tfrac{1}{2} \int \{\varphi_{,A} \delta (\varrho_0 \mathsf{M}_i X_{A,i}) + (\varrho_0 \mathsf{M}_i X_{A,i}) \delta \varphi_{,A}\} d\tau_0. \tag{A.25}$$

This could have been obtained directly from the formula

$$\left. \begin{aligned} W_{\mathrm{m}} &= -\tfrac{1}{2} \int \boldsymbol{M} \cdot \boldsymbol{H}' \, d\tau = \tfrac{1}{2} \int \boldsymbol{M} \cdot \nabla\varphi \, d\tau \\ &= \tfrac{1}{2} \int \varrho \mathsf{M}_i \varphi_{,i} \, d\tau = \tfrac{1}{2} \int \varrho_0 \mathsf{M}_i \varphi_{,A} X_{A,i} \, d\tau_0. \end{aligned} \right\} \tag{A.26}$$

For our purposes, however, it is the separation into two terms that is important:

$$\delta W_{\mathrm{m}} = \delta' W_{\mathrm{m}} + \delta'' W_{\mathrm{m}}, \tag{A.27}$$

where

$$\delta' W_{\mathrm{m}} = \tfrac{1}{2} \int \varphi_{,A} \varrho_0 \delta (\mathsf{M}_i X_{A,i}) d\tau_0 \tag{A.28}$$

and

$$\delta'' W_{\mathrm{m}} = \tfrac{1}{2} \int \varrho_0 \mathsf{M}_i X_{A,i} \delta \varphi_{,A} \, d\tau_0. \tag{A.29}$$

Our problem is to eliminate the variations $\delta\varphi_{,A}$, which are not independent of those of M_i and x_i. We can do this by relating $\delta'' W_{\mathrm{m}}$ to $\delta' W_{\mathrm{m}}$. When there are no strains, $\delta'' W_{\mathrm{m}} = \delta' W_{\mathrm{m}}$; the difference, therefore, must involve only the variations δx_i and not the variations $\delta \mathsf{M}_i$.

By tracing the steps from Eq. (A.25) back to Eq. (A.13) separately for the two terms, we find that

$$\delta' W_{\mathrm{m}} = \frac{1}{2\gamma} \int \varphi_{,A} \delta (\varphi_{,B} X_{A,i} X_{B,i} J) d\tau_0 \tag{A.30}$$

and

$$\delta'' W_{\mathrm{m}} = \frac{1}{2\gamma} \int (\varphi_{,B} X_{A,i} X_{B,i} J) \, \delta \varphi_{,A} \, d\tau_0 . \tag{A.31}$$

Therefore

$$\left. \begin{aligned} \delta' W_{\mathrm{m}} &= \frac{1}{2\gamma} \int \varphi_{,A} [X_{A,i} X_{B,i} J \, \delta \varphi_{,B} + \varphi_{,B} \, \delta (X_{A,i} X_{B,i} J)] \, d\tau_0 \\ &= \delta'' W_{\mathrm{m}} + \frac{1}{2\gamma} \int \varphi_{,A} \, \varphi_{,B} \, \delta (X_{A,i} X_{B,i} J) \, d\tau_0 , \end{aligned} \right\} \tag{A.32}$$

and

$$\delta W_{\mathrm{m}} = \delta' W_{\mathrm{m}} + \delta'' W_{\mathrm{m}} = 2 \delta' W_{\mathrm{m}} - (\delta' W_{\mathrm{m}} - \delta'' W_{\mathrm{m}}) = 2 \delta' W_{\mathrm{m}} + \delta''' W_{\mathrm{m}} , \tag{A.33}$$

with

$$\delta''' W_{\mathrm{m}} = - \frac{1}{2\gamma} \int \varphi_{,A} \, \varphi_{,B} \, \delta (X_{A,i} X_{B,i} J) \, d\tau_0 . \tag{A.34}$$

From Eq. (A.28),

$$\left. \begin{aligned} 2 \delta' W_{\mathrm{m}} &= \int \varphi_{,A} \, \varrho_0 (X_{A,i} \, \delta \mathsf{M}_i + \mathsf{M}_i \, \delta X_{A,i}) \, d\tau_0 \\ &= \int \varphi_{,A} \, \varrho_0 \, X_{A,i} \, \delta \mathsf{M}_i \, d\tau_0 + \int \varphi_{,A} \, \varrho_0 \, \mathsf{M}_i \, \delta X_{A,i} \, d\tau_0 . \end{aligned} \right\} \tag{A.35}$$

From Eq. (A.34),

$$\left. \begin{aligned} \delta''' W_{\mathrm{m}} &= - \frac{1}{2\gamma} \int \varphi_{,A} \, \varphi_{,B} (X_{A,i} \, \delta X_{B,i} J + X_{B,i} \, \delta X_{A,i} J + \\ &\quad + X_{A,i} X_{B,i} \, \delta J) \, d\tau_0 \\ &= - \frac{1}{\gamma} \int \varphi_{,A} \, \varphi_{,B} \, J X_{B,i} \, \delta X_{A,i} \, d\tau_0 - \\ &\quad - \frac{1}{2\gamma} \int \varphi_{,A} \, \varphi_{,B} \, X_{A,i} X_{B,i} \, \delta J \, d\tau_0 . \end{aligned} \right\} \tag{A.36}$$

Substitution of Eqs. (A.35) and (A.36) in Eq. (A.33) gives

$$\left. \begin{aligned} \delta W_{\mathrm{m}} &= \int \varphi_{,A} \, \varrho_0 \, X_{A,i} \, \delta \mathsf{M}_i \, d\tau_0 + \\ &\quad + \frac{1}{\gamma} \int \varphi_{,A} \, \delta X_{A,i} (\gamma \varrho_0 \mathsf{M}_i - \varphi_{,B} J X_{B,i}) \, d\tau_0 - \\ &\quad - \frac{1}{2\gamma} \int \varphi_{,A} \, \varphi_{,B} \, X_{A,i} X_{B,i} \, \delta J \, d\tau_0 . \end{aligned} \right\} \tag{A.37}$$

With use of Eqs. (A.4) and (A.5), this becomes

$$\left. \begin{aligned} \delta W_{\mathrm{m}} &= \int \varphi_{,A} \, \varrho_0 \, X_{A,i} \, \delta \mathsf{M}_i \, d\tau_0 + \\ &\quad + \frac{1}{\gamma} \int \varphi_{,A} (- X_{A,k} X_{C,i} \, \delta x_{k,C}) (\gamma \varrho_0 \mathsf{M}_i - \varphi_{,B} J X_{B,i}) \, d\tau_0 - \\ &\quad - \frac{1}{2\gamma} \int \varphi_{,A} \, \varphi_{,B} \, X_{A,i} X_{B,i} J X_{C,k} \, \delta x_{k,C} \, d\tau_0 \\ &= \int \varphi_{,A} \, \varrho_0 \, X_{A,i} \, \delta \mathsf{M}_i \, d\tau_0 + \int S_{kC} \, \delta x_{k,C} \, d\tau_0 , \end{aligned} \right\} \tag{A.38}$$

with

$$\left.\begin{aligned} S_{kC} = \frac{1}{\gamma}\,\varphi_{,A}(\varphi_{,B}\,JX_{B,i} - \gamma\,\varrho_0\,\mathsf{M}_i)\,X_{A,k}\,X_{C,i} - \\ - \frac{1}{2\gamma}\,\varphi_{,A}\,\varphi_{,B}\,X_{A,i}\,X_{B,i}\,JX_{C,k}. \end{aligned}\right\} \qquad (A.39)$$

The usual integration by parts in the second term of Eq. (A.38) eliminates $\delta x_{k,C}$ and gives

$$\left.\begin{aligned} \delta W_{\mathrm{m}} = \int \varphi_{,A}\,\varrho_0\,X_{A,i}\,\delta\mathsf{M}_i\,d\tau_0 + \int N_C\,S_{kC}\big|_+^-\,\delta x_k\,dS_0 - \\ - \int S_{kC,C}\,\delta x_k\,d\tau_0. \end{aligned}\right\} \qquad (A.40)$$

It remains only to simplify $S_{kC}\big|_+^-$ and $S_{kC,C}$.

On the surface, we have

$$\left.\begin{aligned} N_C\,S_{kC}\,dS_0 = N_C\,dS_0\cdot\frac{1}{\gamma}\,\varphi_{,A}(\varphi_{,B}\,JX_{B,i} - \gamma\,\varrho_0\,\mathsf{M}_i)\,X_{A,k}\,X_{C,i} - \\ - N_C\,dS_0\cdot\frac{1}{2\gamma}\,\varphi_{,A}\,\varphi_{,B}\,X_{A,i}\,X_{B,i}\,JX_{C,k}, \end{aligned}\right\} \qquad (A.41)$$

or by use of Eq. (A.1)

$$\left.\begin{aligned} N_C\,S_{kC}\,dS_0 &= J^{-1}\,n_i\,dS\cdot\frac{1}{\gamma}\,\varphi_{,A}(\varphi_{,B}\,JX_{B,i} - \gamma\,\varrho_0\,\mathsf{M}_i)\,X_{A,k} - \\ &\quad - n_k\,dS\cdot\frac{1}{2\gamma}\,\varphi_{,A}\,\varphi_{,B}\,X_{A,i}\,X_{B,i} \\ &= \frac{1}{\gamma}\,n_i\,dS\,\varphi_{,k}(\varphi_{,i} - \gamma\,\varrho\,\mathsf{M}_i) - \frac{1}{2\gamma}\,n_k\,dS\,\varphi_{,i}\,\varphi_{,i} \\ &= t_{\mathrm{MS}\,k\,i}\,n_i\,dS, \end{aligned}\right\} \qquad (A.42)$$

with

$$\left.\begin{aligned} t_{\mathrm{MS}\,k\,i} &= \frac{1}{\gamma}\left\{\varphi_{,k}\,\varphi_{,i} - \gamma\,\varphi_{,k}\,M_i - \frac{1}{2}\,\delta_{ki}\,\varphi_{,l}\,\varphi_{,l}\right\} \\ &= \frac{1}{\gamma}\left\{H'_k\,H'_i + \gamma\,H'_k\,M_i - \frac{1}{2}\,\delta_{ki}\,H'^2\right\} \\ &= \frac{1}{\gamma}\left(H'_k\,H'_i - \frac{1}{2}\,H'^2\,\delta_{ki}\right) + H'_k\,M_i \end{aligned}\right\} \qquad (A.43)$$

(the notation t_{MS}, for "Maxwell stress", is TOUPIN'S). The discontinuity across S is

$$t_{\mathrm{MS}\,ij}\big|_-^+\,n_j = \tfrac{1}{2}\,\gamma\,M_{\mathrm{n}}^2\,n_i. \qquad (A.44)$$

Hence

$$\left.\begin{aligned} \int N_C\,S_{kC}\big|_+^-\,\delta x_k\,dS_0 &= \int t_{\mathrm{MS}\,k\,i}\big|_+^-\,n_i\,\delta x_k\,dS \\ &= -\tfrac{1}{2}\,\gamma\int M_{\mathrm{n}}^2\,n_k\,\delta x_k\,dS. \end{aligned}\right\} \qquad (A.45)$$

In the interior and exterior regions, we have

$$\left.\begin{aligned} S_{kC,C} = \frac{1}{\gamma}\,\{\varphi_{,A}(\varphi_{,B}\,JX_{B,i} - \gamma\,\varrho_0\,\mathsf{M}_i)\,X_{A,k}\,X_{C,i}\}_{,C} - \\ - \frac{1}{2\gamma}\,(\varphi_{,A}\,\varphi_{,B}\,X_{A,i}\,X_{B,i}\,JX_{C,k})_{,C}, \end{aligned}\right\} \qquad (A.46)$$

or with use of Eq. (A.2)

$$\left.\begin{aligned} S_{kC,C} &= \frac{1}{\gamma} J X_{C,i} \{\varphi_{,A}\, \varphi_{,B}\, X_{B,i}\, X_{A,k} - \gamma J^{-1} \varphi_{,A}\, \varrho_0\, \mathsf{M}_i\, X_{A,k}\}_{,C} - \\ &\quad - \frac{1}{2\gamma} J X_{C,k} (\varphi_{,A}\, \varphi_{,B}\, X_{A,i}\, X_{B,i})_{,C} \\ &= \frac{1}{\gamma} J \{\varphi_{,i}\, \varphi_{,k} - \gamma\, \varphi_{,k}\, M_i\}_{,i} - \frac{1}{2\gamma} J\, (\varphi_{,i}\, \varphi_{,i})_{,k} \\ &= J \cdot \frac{1}{\gamma} \{\varphi_{,k}\, \varphi_{,i} - \gamma\, \varphi_{,k}\, M_i - \frac{1}{2} \delta_{ki}\, \varphi_{,l}\, \varphi_{,l}\}_{,i} = J t_{\mathrm{MS}ki,i}. \end{aligned}\right\} \quad \text{(A.47)}$$

On carrying out the differentiations and using the relations

$$H'_{i,k} = H'_{k,i} \text{ (since } \nabla \times \boldsymbol{H}' = 0) \quad \text{and} \quad (H'_i + \gamma M_i)_{,i} = B'_{i,i} = \nabla \cdot \boldsymbol{B}' = 0,$$

we get

$$S_{kC,C} = J M_i\, H'_{k,i}. \quad \text{(A.48)}$$

The first term in Eq. (A.40) is

$$\int \varphi_{,A}\, \varrho_0\, X_{A,i}\, \delta \mathsf{M}_i\, d\tau_0 = \int \varphi_{,i}\, \varrho\, \delta \mathsf{M}_i\, d\tau = -\int H'_i\, \delta \mathsf{M}_i\, dm. \quad \text{(A.49)}$$

Thus, finally,

$$\left.\begin{aligned} \delta W_{\mathrm{m}} &= -\int H'_i\, \delta \mathsf{M}_i\, dm - \int M_i\, H'_{k,i}\, \delta x_k\, d\tau - \tfrac{1}{2} \gamma \int M_{\mathrm{n}}^2\, n_k\, \delta x_k\, dS \\ &= -\int \boldsymbol{H}' \cdot \delta \boldsymbol{\mathsf{M}}\, dm - \int (\boldsymbol{M} \cdot \nabla \boldsymbol{H}') \cdot \delta \boldsymbol{x}\, d\tau - \tfrac{1}{2} \gamma \int M_{\mathrm{n}}^2\, \boldsymbol{n} \cdot \delta \boldsymbol{x}\, dS. \end{aligned}\right\} \quad \text{(A.50)}$$

Appendix B

Proof of the Magnetic Reciprocity Theorem (11.32)

Let $\boldsymbol{M}_1$ and $\boldsymbol{M}_2$ be continuous within the volume τ inside closed surface S; let $\boldsymbol{M}_1 = \boldsymbol{M}_2 = 0$ thruout the volume τ' outside S. Then $\boldsymbol{H}_1$ and $\boldsymbol{H}_2$ are regular at infinity, and

$$\int_\tau \left[\boldsymbol{H}_1 \cdot \frac{\partial \boldsymbol{H}_2}{\partial z} + \boldsymbol{H}_2 \cdot \frac{\partial \boldsymbol{H}_1}{\partial z}\right] d\tau = \int_\tau \frac{\partial}{\partial z} [\boldsymbol{H}_1 \cdot \boldsymbol{H}_2]\, d\tau = \int_S \boldsymbol{H}_1^- \cdot \boldsymbol{H}_2^-\, n_3\, dS, \quad \text{(B.1)}$$

$$\int_{\tau'} \left[\boldsymbol{H}_1 \cdot \frac{\partial \boldsymbol{H}_2}{\partial z} + \boldsymbol{H}_2 \cdot \frac{\partial \boldsymbol{H}_1}{\partial z}\right] d\tau = -\int_S \boldsymbol{H}_1^+ \cdot \boldsymbol{H}_2^+\, n_3\, dS, \quad \text{(B.2)}$$

where superscripts $^+$ and $^-$ denote the outer and inner values, respectively, on S. Addition gives

$$\int_{\text{space}} \left[\boldsymbol{H}_1 \cdot \frac{\partial \boldsymbol{H}_2}{\partial z} + \boldsymbol{H}_2 \cdot \frac{\partial \boldsymbol{H}_1}{\partial z}\right] d\tau = -\int_S (\boldsymbol{H}_1 \cdot \boldsymbol{H}_2)\big|_-^+\, n_3\, dS, \quad \text{(B.3)}$$

where $|_-^+$ means the outer value minus the inner. We have also, if φ_i is the potential corresponding to $\boldsymbol{H}_i$,

$$\left.\begin{aligned}\int_{\text{space}} \boldsymbol{B}_1 \cdot \frac{\partial \boldsymbol{H}_2}{\partial z}\, d\tau &= -\int \boldsymbol{B}_1 \cdot \nabla \frac{\partial \varphi_2}{\partial z}\, d\tau \\ &= -\int_{\text{space}} [\nabla \cdot (\boldsymbol{B}_1\, \partial\varphi_2/\partial z) - (\nabla \cdot \boldsymbol{B}_1)\, \partial\varphi_2/\partial z]\, d\tau \\ &= +\int_S \boldsymbol{n} \cdot \boldsymbol{B}_1 (\partial\varphi_2/\partial z)|_-^+\, dS = -\int_S \boldsymbol{n} \cdot \boldsymbol{B}_1\, \boldsymbol{k} \cdot (\boldsymbol{H}_2|_+^-)\, dS,\end{aligned}\right\} \tag{B.4}$$

since $\nabla \cdot \boldsymbol{B}_1 = 0$ and $\boldsymbol{n} \cdot \boldsymbol{B}_1$ is continuous across S. On adding to Eq. (B.4) the corresponding equation with 1 and 2 interchanged, we get

$$\left.\begin{aligned}\int_{\text{space}} \left[\boldsymbol{B}_1 \cdot \frac{\partial \boldsymbol{H}_2}{\partial z} + \boldsymbol{B}_2 \cdot \frac{\partial \boldsymbol{H}_1}{\partial z}\right] d\tau \\ = -\int_S [\boldsymbol{n} \cdot \boldsymbol{B}_1\, \boldsymbol{k} \cdot (\boldsymbol{H}_2|_-^+) + \boldsymbol{n} \cdot \boldsymbol{B}_2\, \boldsymbol{k} \cdot (\boldsymbol{H}_1|_-^+)]\, dS.\end{aligned}\right\} \tag{B.5}$$

Subtraction of (B.3) from (B.5) gives, since $\boldsymbol{B}_i - \boldsymbol{H}_i = \gamma \boldsymbol{M}_i$ and since $\boldsymbol{M}_i = 0$ in τ',

$$\left.\begin{aligned}\int_\tau \left[\boldsymbol{M}_1 \cdot \frac{\partial \boldsymbol{H}_2}{\partial z} + \boldsymbol{M}_2 \cdot \frac{\partial \boldsymbol{H}_1}{\partial z}\right] d\tau \\ = \frac{1}{\gamma} \int_S \{-\boldsymbol{n} \cdot \boldsymbol{B}_1\, \boldsymbol{k} \cdot (\boldsymbol{H}_2|_-^+) - \boldsymbol{n} \cdot \boldsymbol{B}_2\, \boldsymbol{k} \cdot (\boldsymbol{H}_1|_-^+) + \\ + (\boldsymbol{H}_1 \cdot \boldsymbol{H}_2)|_-^+\, n_3\}\, dS.\end{aligned}\right\} \tag{B.6}$$

Now

$$\boldsymbol{H}_i|_-^+ = \boldsymbol{n}\, H_{in}|_-^+ = \gamma \boldsymbol{n} M_{in} \quad (i = 1, 2), \tag{B.7}$$

hence

$$\boldsymbol{k} \cdot (\boldsymbol{H}_i|_-^+) = \gamma\, n_3\, M_{in}; \tag{B.8}$$

also

$$(\boldsymbol{H}_1 \cdot \boldsymbol{H}_2)|_-^+ = (H_{1n} H_{2n} + \boldsymbol{H}_{1t} \cdot \boldsymbol{H}_{2t})|_-^+ = (H_{1n} H_{2n})|_-^+, \tag{B.9}$$

since the tangential part $\boldsymbol{H}_{it}$ of $\boldsymbol{H}_i$ is continuous across S. If $\overline{\boldsymbol{H}}$ represents the average of the inside and outside values,

$$\left.\begin{aligned}(\boldsymbol{H}_1 \cdot \boldsymbol{H}_2)|_-^+ &= (\overline{H}_{1n} + \tfrac{1}{2}\gamma M_{1n})(\overline{H}_{2n} + \tfrac{1}{2}\gamma M_{2n}) - \\ &\quad - (\overline{H}_{1n} - \tfrac{1}{2}\gamma M_{1n})(\overline{H}_{2n} - \tfrac{1}{2}\gamma M_{2n}) \\ &= \gamma (M_{1n} \overline{H}_{2n} + M_{2n} \overline{H}_{1n}).\end{aligned}\right\} \tag{B.10}$$

On using (B.8) and (B.10), we find that the surface integrand in Eq. (B.6) reduces to

$$\left.\begin{aligned}& n_3 \{M_{2n}(-B_{1n} + \overline{H}_{1n}) + M_{1n}(-B_{2n} + \overline{H}_{2n})\} \\ &= n_3 \{-M_{2n} \cdot \gamma \overline{M}_{1n} - M_{1n} \cdot \gamma \overline{M}_{2n}\} \\ &= -\tfrac{1}{2}\gamma\, n_3 \{M_{2n} M_{1n} + M_{1n} M_{2n}\} = -\gamma\, n_3\, M_{1n} M_{2n},\end{aligned}\right\} \tag{B.11}$$

since $\overline{M}_n = \frac{1}{2}(M_n + 0) = \frac{1}{2}M_n$. Hence

$$\int_\tau \left[\boldsymbol{M}_1 \cdot \frac{\partial \boldsymbol{H}_2}{\partial z} + \boldsymbol{M}_2 \cdot \frac{\partial \boldsymbol{H}_1}{\partial z}\right] d\tau = -\gamma \int_S n_3 \, M_{1n} \, M_{2n} \, dS, \qquad \text{(B.12)}$$

which is Eq. (11.32). Obviously the direction here chosen as the z axis can be any fixed direction in space.

Appendix C

On Angular Velocity

As all students of elementary hydrodynamics know, the velocity gradient tensor $v_{i,j}$ at a point P of a fluid or of a deformable solid can be separated into a symmetric part $v_{(i,j)}$ and an antisymmetric part $v_{[i,j]}$, of which the former describes a "rate of strain" and the latter a "rigid-body rotation" of the matter in the neighborhood of P. The antisymmetric tensor $v_{[i,j]}$ is equivalent to an axial vector $\boldsymbol{\Omega}$ with components

$$\Omega_1 = \tfrac{1}{2}(v_{3,2} - v_{2,3}) = v_{[3,2]} = -v_{[2,3]}, \ldots \qquad \text{(C.1)}$$

(where the differentiation is with respect to the deformed coordinates x_i). It is natural to regard $\boldsymbol{\Omega}$ as representing "the" angular velocity of the matter in a neighborhood of P. The purpose of this appendix is to demonstrate, by an example, that this natural interpretation of $\boldsymbol{\Omega}$ is wrong, except as an approximation within the limitations of infinitesimal-strain theory (or, of course, when the motion in a neighborhood of P is actually rigid).

Let the undeformed coordinates be

$$X_1 = R\cos\Theta, \qquad X_2 = R\sin\Theta, \qquad X_3 \qquad \text{(C.2)}$$

and let the deformed coordinates be

$$x_1 = K_1 R\cos(\Theta + \omega t), \qquad x_2 = K_2 R\sin(\Theta + \omega t), \qquad x_3 = X_3, \qquad \text{(C.3)}$$

where K_1, K_2, and ω are constants, with K_1 and K_2 positive and unequal. The components of velocity are then

$$\left.\begin{aligned} v_1 &= \dot{x}_1 = -\omega K_1 R\sin(\Theta + \omega t), \\ v_2 &= \dot{x}_2 = \omega K_2 R\cos(\Theta + \omega t), \\ v_3 &= \dot{x}_3 = 0; \end{aligned}\right\} \qquad \text{(C.4)}$$

on elimination of R and Θ by use of Eqs. (C.3), we get

$$v_1 = -\omega\frac{K_1}{K_2}x_2, \qquad v_2 = \omega\frac{K_2}{K_1}x_1, \qquad v_3 = 0 \qquad \text{(C.5)}$$

and hence

$$\Omega_1 = \Omega_2 = 0, \quad \Omega_3 = \frac{1}{2}\left(\frac{\partial v_2}{\partial x_1} - \frac{\partial v_1}{\partial x_2}\right) = \frac{1}{2}\,\omega\left(\frac{K_2}{K_1} + \frac{K_1}{K_2}\right) \equiv \Omega. \tag{C.6}$$

For a deformable solid, these equations describe a certain possible rotation of a naturally circular cylinder in a rigid cylindrical hole of elliptic cross section. For a fluid, we may start with Eqs. (C.5); by integration, we obtain equations equivalent to Eqs. (C.3) and conclude that each particle (except those on the x_3 axis) moves in an elliptic path. In either case, the period of the motion is $2\pi/\omega$. If, therefore, there is any quantity deserving of the name "the angular velocity of the matter about O", that quantity is surely $\boldsymbol{\omega} = \omega \boldsymbol{k}$. Eq. (C.1), however, give an "angular velocity" $\boldsymbol{\Omega} = \Omega \boldsymbol{k}$ that differs from $\boldsymbol{\omega}$ by a factor

$$\frac{\Omega}{\omega} = \frac{1}{2}\left(\frac{K_2}{K_1} + \frac{K_1}{K_2}\right). \tag{C.7}$$

This attains its minimum value 1 when $K_2/K_1 = 1$ and exceeds 1 when $K_2/K_1 \neq 1$.

If K_1 and K_2 differ from 1 by small quantities ε_1 and ε_2 respectively, then to the second order in ε_1 and ε_2

$$\frac{\Omega}{\omega} = 1 + \frac{1}{2}(\varepsilon_1 - \varepsilon_2)^2. \tag{C.8}$$

Since there is no first-order term, identification of Ω with ω is correct within the approximations of linear elasticity theory. It is incorrect in the second order of small quantities.

The finite rotation tensor $R_{A\,i}$ is

$$R_{A\,i} = \begin{bmatrix} \cos\omega t & -\sin\omega t & 0 \\ \sin\omega t & \cos\omega t & 0 \\ 0 & 0 & 1 \end{bmatrix} \tag{C.9}$$

(A numbers the rows, i the columns). This represents a rotation thru angle ωt about the X_3 axis. Thus the rate of change of the finite rotation angle is ω, not Ω.

References

AHARONI, A.: [1] J. Appl. Phys. **34**, 2434 (1963).

— [2] Proc. Internatl. Conf. on Magnetism, Nottingham, 7th—11th September 1964, p. 686—688. London: The Institute of Physics and The Physical Society 1965.

—, and S. SHTRIKMAN: [1] Phys. Rev. **109**, 1522—1528 (1958).

AKULOV, N. S.: [1] Z. Physik **52**, 389—405 (1928).

— [2] Z. Physik **57**, 249—256 (1929).

BATES, L. F.: [1] Modern magnetism (3d ed.). London: Cambridge University Press 1951.

BECKER, R.: [1] Z. Physik **62**, 253—269 (1930).

—, u. W. DÖRING: [1] Ferromagnetismus. Berlin: Springer 1939.

BOZORTH, R. M.: [1] Ferromagnetism. Princeton, New Jersey: D. van Nostrand Co. 1951.

BRILLOUIN, L.: [1] Science and information theory. New York: Academic Press 1956.

BROWN jr., W. F.: [1] J. Appl. Phys. **11**, 160—172 (1940).

— [2] Am. J. Phys. **8**, 338—345 (1940).

— [3] Phys. Rev. **58**, 736—743 (1940).

— [4] Phys. Rev. **60**, 139—147 (1941).

— [5] Am. J. Phys. **19**, 290—304 and 333—350 (1951).

— [6] Revs. Modern Phys. **25**, 131—135 (1953).

— [7] J. Appl. Phys. **29**, 470—471 (1958).

— [8] J. Phys. Soc. Japan **17**, Suppl. BI, 540—542 (1962).

— [9] J. Appl. Phys. **33**, 3026—3031 (1962).

— [10] Magnetostatic principles in ferromagnetism. Amsterdam: North-Holland Publ. Co. 1962.

— [11] Micromagnetics. New York: John Wiley & Sons (Interscience) 1963.

— [12] J. Appl. Phys. **35**, 2102—2106 (1964).

— [13] J. Appl. Phys. **36**, 994—1000 (1965).

— [14] Proc. Internatl. Conf. on Magnetism, Nottingham, 7th—11th September 1964, p. 681—685. London: The Institute of Physics and The Physical Society 1965.

—, and A. H. MORRISH: [1] Phys. Rev. **105**, 1198—1201 (1957).

—, and S. SHTRIKMAN: [1] Phys. Rev. **125**, 825—828 (1962).

CAUCHY, A. L.: [1] Oeuvres, Ser. I, **2**, 356—361.

CHIKAZUMI, S.: [1] Physics of magnetism. New York: John Wiley & Sons 1963.

Coulomb's Law Committee: [1] Am. J. Phys. **18**, 1—25 and 69—88 (1950).

CRAIK, D. J., and R. S. TEBBLE: [1] Repts. Progr. Phys. **24**, 116—166 (1961).

— [2] Ferromagnetism and ferromagnetic domains. Amsterdam: North-Holland Publ. Co. 1965.

DEVONS, S.: [1] Sci Progr. **51**, 601—610 (1963).

EASTMAN, D. E.: [1] Thesis, Mass. Inst. Tech., 1965.

— [2] J. Appl. Phys. **37**, 996—997 (1966).

FELDTKELLER, E.: [1] Z. angew. Phys. **19**, 530—536 (1965).

FISHER, R. A.: [1] J. Roy. Statistical Soc. **98**, pt. I, 39—54 (1935).

GERSDORF, R.: [1] Physica **26**, 553—574 (1960).

HAYASI, T.: [1] Z. Physik **72**, 177—190 (1931); **91**, 818—819 (1934).

HILBERT, D., and W. ACKERMANN: [1] Principles of mathematical logic. Chelsea, New York 1950. [Original: Grundzüge der theoretischen Logik. Berlin: Springer 1938.]

KITTEL, C.: [1] Revs. Modern Phys. **21**, 541—583 (1949).

—, and J. K. GALT: [1] Solid State Phys. **3**, 437—564 (1956).

Kneller, E.: [1] Ferromagnetismus. Berlin-Göttingen-Heidelberg: Springer 1962.
Kornetzki, M.: [1] Z. Physik **87**, 560—579 (1934).
Kronmüller, H.: [1] Magnetisierungskurve der Ferromagnetika. I. In: Moderne Probleme der Metallphysik (A. Seeger, editor), vol. 2. Berlin: Springer 1966.
Landau, L., and E. Lifshitz: [1] Physik. Z. Sowjetunion **8**, 153—169 (1935).
— — [2] Electrodynamics of continuous media. London: Pergamon Press; Reading, Mass.: Addison-Wesley 1960.
Leathem, J. G.: [1] Volume and surface integrals used in physics. 2d ed. London: Cambridge University Press 1913.
Lee, E. W.: [1] Repts. Progr. Phys. **18**, 184—229 (1955).
Lifshitz, E. [1] J. Phys. U.S.S.R. **8**, 337—345 (1944).
Lorentz, H. A.: [1] Theory of electrons. Leipzig: Teubner 1909; New York: Dover 1952.
Love, A. E. H.: [1] A treatise on the mathematical theory of elasticity, 4th ed. London: Cambridge University Press 1934.
Mason, M., and W. Weaver: [1] The electromagnetic field. Chicago: Chicago University Press 1929.
Mason, W. P.: [1] Phys. Rev. **82**, 715—723 (1951).
Maxwell, J. C.: [1] A treatise on electricity and magnetism, 3d ed., vols. I and II. London: Oxford University Press 1904.
Morrish, A. H.: [1] The physical principles of magnetism. New York: John Wiley & Sons 1965.
Muller, M. W., and A. Wehlau: [1] J. Appl. Phys. **32**, 2448—2450 (1961).
Néel, L.: [1] J. phys. radium **15**, 225—239 (1954).
Powell, F. C.: [1] Proc. Cambridge Phil. Soc. **27**, 561—569 (1931).
Seeger, A., and H. Kronmüller: [1] J. Phys. Chem. Solids **12**, 298—313 (1960); **18**, 93—115 (1961).
Shannon, C.: [1] Bell system Tech. J. **27**, 379—423 and 623—656 (1948).
Shtrikman, S., and D. Treves: [1] Micromagnetics. In: Magnetism, a treatise on modern theory and materials (G. T. Rado and H. Suhl, editors), vol. III. New York: Academic Press 1963.
Sokolnikoff, I. S.: [1] Mathematical theory of elasticity. New York: McGraw-Hill Book Co. 1946.
Stewart, K. H.: [1] Ferromagnetic domains. London: Cambridge University Press 1954.
Stoner, E. C.: [1] Phil. Mag. **36**, 803—821 (1945).
—, and E. P. Wohlfarth: [1] Phil. Trans. Roy. Soc. London A **240**, 599—644 (1948).
Tiersten, H. F.: [1] J. Math. and Phys. **5**, 1—21 (1964).
— [2] J. Math. and Phys. **6**, 779—787 (1965).
Toupin, R.: [1] J. Rational Mechanics and Analysis **5**, 850—915 (1956).
Träuble, H.: [1] Magnetisierungskurve der Ferromagnetika. II. In: Moderne Probleme der Metallphysik (A. Seeger, editor), vol. 2. Berlin: Springer 1966.
Truesdell, C.: [1] J. Rational Mechanics and Analysis **1**, 125—300 (1952).
—, and R. A. Toupin: [1] In: Encyclopedia of physics, vol. III/1. Berlin-Göttingen-Heidelberg: Springer 1960.
Turov, E. A.: [1] Physical properties of magnetically ordered crystals. New York: Academic Press 1965. [Original: Э. А. Туров, Физические Свойства Магнитоулорядоченных Кристаллов. Изд. Акад. Наук СССР, Москва, 1963.]
Voigt, W.: [1] Lehrbuch der Kristallphysik. Leipzig: Teubner 1928.
Wohlfarth, E. P.: [1] Advances in Phys. **8**, 87—224 (1959).
— [2] In: Magnetism (G. T. Rado and H. Suhl, editors), vol. III, chap. 7. New York: Academic Press 1963.

Author Index

This index covers only the text, pages 1—145, and not the bibliography, pages 146—147. Joint articles are cited under each of the authors.

Subject Index

Springer Tracts in Natural Philosophy

Edited by C. Truesdell
Co-Editors: L. Collatz, G. Fichera, P. Germain, J. Keller, A. Seeger

---

Vol. 1 **Linearized Analysis of One-Dimensional Magnetohydrodynamic Flows**
By Dr. Roy M. Gundersen, Professor of Mathematics, University of Wisconsin-Milwaukee
With 10 figures. X, 119 pages 8vo. 1964. Cloth DM 22,—

Vol. 2 **Differential- und Integral-Ungleichungen**
und ihre Anwendung bei Abschätzungs- und Eindeutigkeitsproblemen
Von Wolfgang Walter, o. Professor der Mathematik an der Technischen Hochschule Karlsruhe
Mit 18 Abbildungen. XIV, 269 Seiten Gr.-8°. 1964. Ganzleinen DM 59,—

Vol. 3 **Konstruktive Methoden der konformen Abbildung**
Von Dr. rer. nat. Dieter Gaier, Ph.D., o. Professor für angewandte Mathematik an der Justus Liebig-Universität Gießen
Mit 20 Abbildungen und 28 Tabellen. XIV, 294 Seiten Gr.-8°. 1964 Ganzleinen DM 68,—

Vol. 4 **Approximation von Funktionen und ihre numerische Behandlung**
Von Dr. Günter Meinardus, o. Professor an der Technischen Hochschule Clausthal
Mit 21 Abbildungen. VIII, 180 Seiten Gr.-8°. 1964. Ganzleinen DM 49,—

Vol. 5 **Viscometric Flows of Non-Newtonian Fluids**
Theory and Experiment
By B. D. Coleman, Senior Fellow, Mellon Institute,
H. Markovitz, Senior Fellow, Mellon Institute and
W. Noll, Professor of Mathematics, Carnegie Institute of Technology
With 37 figures. XII, 130 pages 8vo. 1966. Cloth DM 22,—

Vol. 6 **Studies in Non-Linear Stability Theory**
By Wiktor Eckhaus, Département de Mécanique, Université de Paris
With 12 figures. VIII, 117 pages 8vo. 1965. Cloth DM 22,—

Vol. 7 **The General Problem of the Motion of Coupled Rigid Bodies about a Fixed Point**
By Dr. Eugene Leimanis, Professor of Mathematics, The University of British Columbia, Vancouver, B.C.
With 66 figures. XVI, 337 pages 8vo. 1965. Cloth DM 48,—

Vol. 8 **Vibrations non linéaires et théorie de la stabilité**
Par Maurice Roseau, Professeur à la Faculté des Sciences de Paris
XII, 254 pages 8vo. 1966. Pleine toile DM 39,—
